GENETIC STUDIES IN DIOECIOUS MELANDRIUM

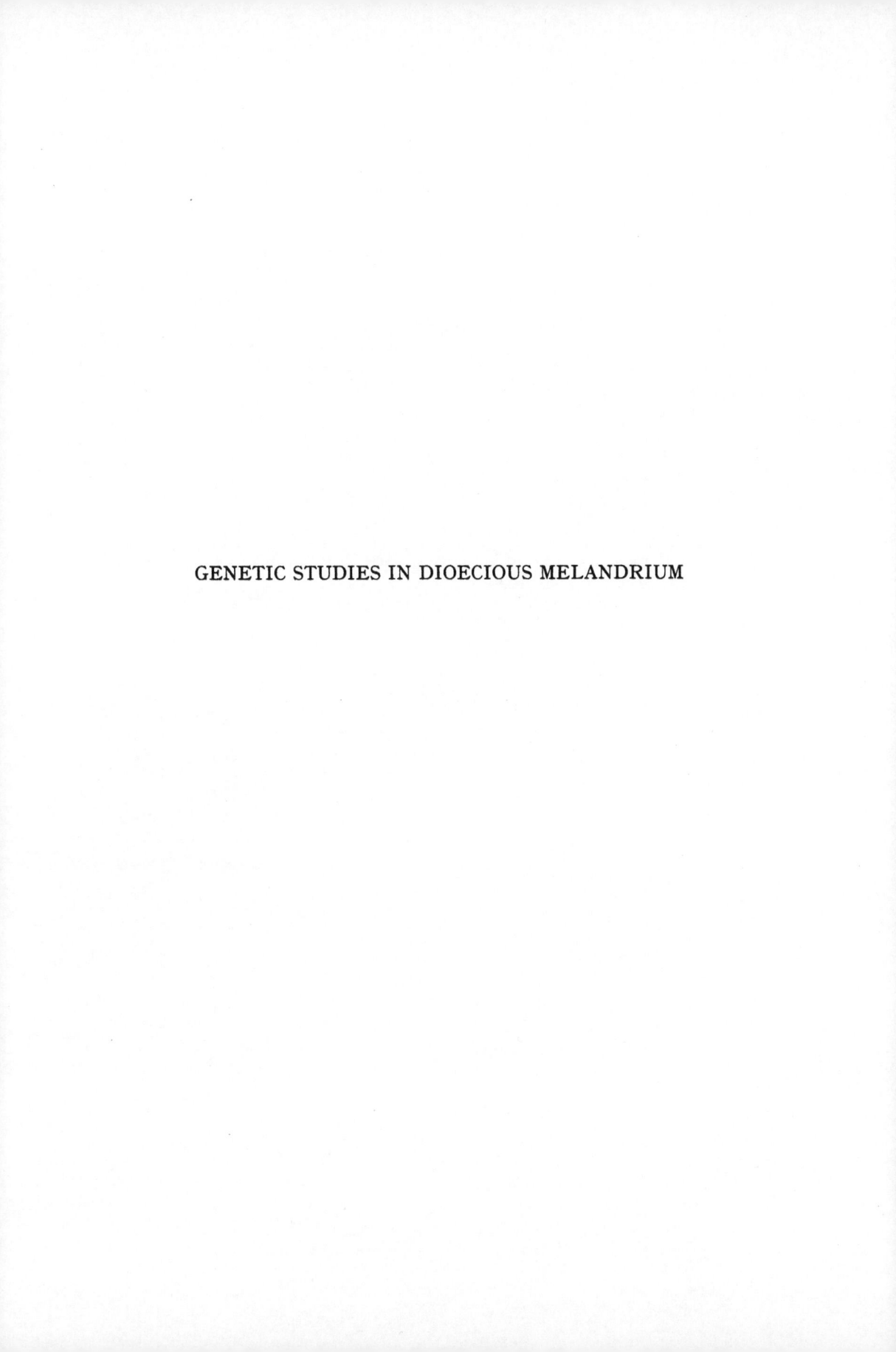

GENETIC STUDIES
IN DIOECIOUS MELANDRIUM

PROEFSCHRIFT

TER VERKRIJGING VAN DE GRAAD VAN DOCTOR
IN DE WISKUNDE EN NATUURWETENSCHAPPEN
AAN DE RIJKSUNIVERSITEIT TE UTRECHT, OP
GEZAG VAN DE RECTOR MAGNIFICUS PROF. DR.
A. R. HULST, VOLGENS BESLUIT VAN DE SENAAT
IN HET OPENBAAR TE VERDEDIGEN OP MAAN-
DAG 3 OCTOBER 1966 DES NAMIDDAGS TE 4 UUR
PRECIES

DOOR

GERRIT VAN NIGTEVECHT

geboren op 15 maart 1930 te Hilversum

SPRINGER-SCIENCE+BUSINESS MEDIA, B.V. 1966

ISBN 978-94-017-6859-7   ISBN 978-94-017-6923-5 (eBook)
DOI 10.1007/978-94-017-6923-5
Promotor: PROF. DR. G. A. VAN ARKEL

Aan mijn Ouders
Aan mijn Vrouw

VOORWOORD

Gaarne maak ik van de gelegenheid gebruik om bij het voltooien van dit proefschrift een woord van dank te richten tot hen die aan de totstandkoming ervan hebben bijgedragen.

Allereerst geldt mijn dank U, Vader en Moeder. U hebt het als vanzelfsprekend beschouwd dat ik biologie zou gaan studeren, ondanks de problemen die dit voor U met zich meebracht. Moge het verschijnen van dit proefschrift U alle voldoening schenken.

In eerbiedige dankbaarheid gedenk ik mijn helaas te vroeg overleden leermeester Prof. Dr. C. L. Rümke, die een zo groot aandeel in het tot stand komen van dit proefschrift heeft gehad. Hij plaatste Melandrium op het programma omdat hij van mening was dat het probleem van de geslachtsdeterminatie nog geruime tijd actueel zou blijven. De in dit proefschrift beschreven experimenten zijn grotendeels onder zijn leiding verricht. Deze leiding was gekenmerkt door de grote mate van vrijheid welke hij zijn medewerkers liet. Het was echter een vrijheid in gebondenheid, doordat hij tegelijk een sterk gevoel van medeverantwoordelijkheid op zijn medewerkers wist over te dragen.

U, Mevrouw Rümke wil ik zeggen dankbaar te zijn dat ik onder zijn leiding heb mogen werken.

Hoogleraren, Lectoren en leden van de wetenschappelijk staf van de Faculteit der Wiskunde en Natuurwetenschappen wil ik gaarne mijn dank betuigen voor de op colleges en practica ontvangen kennis en vorming.

Hooggeleerde Lanjouw, het is voor mij een waardevolle ervaring geweest dat ik enige jaren als assistent aan Uw instituut was verbonden. Ik bewaar de prettigste herinneringen aan die tijd.

Hooggeleerde Jonker, het contact dat ik in de vorengenoemde periode met U mocht hebben, heeft meer invloed op mijn vorming gehad dan ik destijds besefte.

Hooggeleerde van Arkel, Hooggeachte Promotor, beste Gerard, de wijze waarop je na het overlijden van Professor Rümke diens taken hebt overgenomen en uitgevoerd dwingt respect af. Eén van deze taken hield in dat je mijn Promotor zou zijn. Je opbouwende critiek en de vele discussies welke wij hebben gevoerd zijn dit proefschrift zeer ten goede gekomen. Hiervoor ben ik je zeer erkentelijk.

Bijzonder veel dank ben ik verschuldigd aan alle medewerkers van ,,de Proeftuin''. In de beginperiode waren dit de Heren J. J. Viets en J. W. Hofland die op voortreffelijke wijze de technische aspecten van ,,de Proeftuin'' verzorgden. Na pensionering van de Heer Viets werd diens taak overgenomen door de Heer C. L. L. H. van Woerden, die ik zeer erkentelijk ben voor de wijze waarop hij mij van zijn veelzijdige ervaring heeft laten profiteren. De komst van de Heer H. W. de Groot maakt dat de steeds intensiever gebruikte proeftuin naar behoren kan worden verzorgd.

Een bijzonder woord van dank is hier wel op zijn plaats voor de voortreffelijke hulp op velerlei gebied welke ik van jou Petra, en ook van je voorgangster Ineke, heb mogen ontvangen.

Beste Jettie, Weledelgeleerde Vrouwe Nienhuis-van Albada, het enthousiasme waarmee jij je aan het onderzoek wijdt, werkt verkwikkend. Dat je verschillende van mijn taken hebt overgenomen betekent voor mij een welkome verlichting.

Het is onmogelijk de vele studenten die aan het onderzoek hebben medegewerkt met name te noemen. Hun medewerking heb ik steeds als waardevol ervaren.

Alle medewerkers van het Genetisch Instituut zeg ik dank voor de ondervonden hulp en medewerking. In het bijzonder Mejuffrouw E. G. Burgemeestre die, naast haar vele andere werkzaamheden, met Mejuffrouw A. A. van Varik veel tijd besteed heeft aan het typewerk. Onze onvolprezen tekenaar en fotograaf de Heer Dick Smit dank ik voor de keurige verzorging van de illustratie.

Bijzonder erkentelijk ben ik Ir. J. J. Bezem voor de adviezen die ik in verband met enkele statistische bewerkingen van hem mocht ontvangen.

Herma, door de wijze waarop jij mijn geabsorbeerd zijn door het werk weet op te vangen, heb je een groter bijdrage tot de voltooiing van dit proefschrift geleverd, dan ik zou kunnen beschrijven. Heb dank hiervoor.

CONTENTS

Genetica (1966) **37**: 281–306

GENETIC STUDIES IN DIOECIOUS *MELANDRIUM*. I.

SEX-LINKED AND SEX-INFLUENCED INHERITANCE IN *MELANDRIUM ALBUM* AND *MELANDRIUM DIOICUM*

G. VAN NIGTEVECHT

Institute of Genetics, State University, Utrecht, The Netherlands
(*Received April 1, 1966*)

Sex-linked and sex-influenced inheritance are of interest because of their relation to the still intriguing problem of sex determination. Genes involved in the formation of the sex organs are regarded to be sex-determining genes. These genes may be present in all chromosomes including the sex-chromosomes. Other genes present in the sex-chromosomes, but not involved in sex determination, are the sex-linked genes. A mutation for narrow leaves we came across in our *M. album* material is regarded as a case of sex linkage. Also the certation effect observed in *M. album* and *M. dioicum* must have been caused by genes on the sex-chromosomes. In both cases, however, it is not altogether unlikely, that the genes, regarded as sex-linked ones, actually take part in the process of sex-determination.

Sex-determining genes might influence the effect of other genes, that are therefore called sex-influenced genes. We observed a number of such sex-influenced characters in *Melandrium*.

In *M. album*, female plants are, on the whole, larger than male plants, having larger stems and leaves. The petals, however, are larger in male plants, except in families with very broad petals. The leaves and petals are narrower in female plants than in male ones, except in families with very broad leaves and families with broad petals, where the difference in shape was no longer present. Usually, slightly more anthocyanin is formed in male plants than in females both in petals and in green parts. More glandular hairs were observed on male plants than on female ones.

Insofar the observations were made in *M. dioicum* the same results were obtained.

We regard these phenomena to be an expression of the different physiological conditions in female and in male plants, these conditions being provoked by the sex-determining genes and more favourable for vegetative growth in female than in male plants.

Introduction

Already at the beginning of this century, DE VRIES (1900), CORRENS (1900–), BATESON (1902–), BAUR (1910–), SHULL (1910–) and WINGE (1917–) made more or less extensive investigations in *Melandrium*.

Their main reason for undertaking these investigations was that shortly after the rediscovery of Mendel's laws, the genetical determination of sex and sex differences became an important topic. The dioecious *M. album* and *M. dioicum* appeared to be very suitable for this purpose. In both species the female plants have 2 X-chromosomes and 22 autosomes and the male plants an X- and a Y-chromosome and 22 autosomes.

Genes that are situated in the sex chromosomes but which are not involved in sex determination are sex-linked genes. Genes that are not involved in sex determination but whose effect is influenced by the sex-determining mechanism are called sex-influenced genes. They may be present in all chromosomes.

The first case of sex-linkage found in plants, the recessive X-linked gene *angustifolia*, was described by BAUR (1912) and SHULL (1914). WINGE (1927) proved that this gene, responsible for extremely narrow folia in *M. album*, was lethal to pollen. Thus only male plants with such narrow folia could arise.

WINGE (1931) found three other sex-linked genes: *aurea*, *abnormal*, and an inhibitor for *variegated*. He detected all three types in the generations that followed upon crosses between *M. album* and *M. dioicum*.

The gene *aurea* is a recessive gene in the X-chromosome. When it is homozygous it is lethal, therefore no *aurea* female plants are found. According to WINGE the gene is not lethal in the hemizygous condition, so that yellowish-green male plants can occur. In the presence of an autosomal inhibitor *A*, the *aurea* gene is neither lethal nor displays the yellowish-green phenotype.

The recessive gene *abnormal* is localized in the homologous parts of the X- and the Y-chromosome. Three autosomal inhibitors, *G*, *H* and *I* were found. As described by WINGE (1932), the *abnormal* plants have their calyx constantly closed, because of which the flowers never open.

The Y-chromosome contains an inhibitor for the autosomal recessive gene *variegated*, since no *variegated* male individuals occur. Moreover, three autosomal inhibitors *L*, *M* and *O* were detected.

Differences between factors in the X- and Y-chromosome active in the haploid phase were demonstrated by the experiments of CORRENS (1921, 1922, 1924, 1926). He demonstrated certation of gametes which, as we now know, have different sex chromosomes. Pollen tubes of the gametes with an X-chromosome grow more rapidly than the pollen tubes of the gametes with a Y-chromosome.

HARTSHORNE (1963) found no certation in *M. dioicum*, but LAW-RENCE (1963, 1964) demonstrated a certation effect both in *M. album* and in *M. dioicum*.

Gametes with a Y-chromosome are more resistant to storage and to alcoholic vapours than are gametes with an X-chromosome, (CORRENS, 1922, 1924).

A number of investigations in *Melandrium* are known where it is not always clear whether cases of sex-linkage or of sex-influenced inheritance are described.

SCHULZ (1890) stated that in *M. album* female plants have smaller petals than male plants.

LÖVE (1944) and BAKER (1951) observed that female plants of *M. album* have bigger flowers than male individuals, whereas in *M. dioicum* the reverse situation is met with.

LÖVE (1944) measured the length of the calyx, the length of the whole flower and the diameter of the corolla. In all cases the sizes of these parts in the pistillate plants of *M. album* and in the staminate ones of *M. dioicum* exceeded those in the other sex of the same species. LÖVE (1944) assumed this to be due to sex-linked factors. BAKER (1951) measured the length of the leaves on the fifth node on a shoot. Neither in *M. album* nor in *M. dioicum* did he find any differences in the l/w ratios between male and female plants.

According to BAKER (1951) the intensity of the petal colour is greater in male than in female flowers. LÖVE (1944) supposed some sex-linked genes for acidity in the petals. WINGE (1931) obtained some peculiar results after crossing *M. dioicum* × *M. album* and *M. album* × *M. dioicum*. In the first cross the flowers of the male plants were, on the whole, of a deeper red than the female flowers, in the second cross, however, the female flowers were more intensily coloured than the male flowers. According to WINGE this must be due to sex-linked factors.

STANFIELD (1937, 1944) demonstrated differences in the chemical composition of roots and tops between male and female individuals of *M. album*. He found the highest phosphorus and sugar content in both the vegetative and early flowering phases of female plants. In the early flowering stage female plants showed a greater phenoloxidase activity in roots and tops than male plants.

Our investigations were undertaken because several data given in the literature seem to contradict each other. We are of opinion that it is of interest to investigate which characters are controlled by sex-linked genes and which by sex-influenced genes. Sex-influenced phenomena might give an indication about the different physiological conditions in female and in male plants which cause the development of respectively gynoecia and androecia. We therefore assume sex-influenced characters to be of importance for obtaining a better insight into the mechanism of sex determination in *M. album* and *M. dioicum*.

Material and Methods

Melandrium album (Mill.) Garcke and *M. dioicum* (L.emend.) Coss. et Germ. are two dioecious species, sometimes regarded as subspecies (LÖVE, 1944), of the Caryophyllaceae. In "Flora europaea" (TUTIN, HEYWOOD et al., 1964) the species are described as *Silene alba* (Miller) E. H. L. Krause and *Silene dioica* (L.) Clairv.

A comprehensive description of both species and their ecotypes has been given bij BAKER (1947, 1948). LÖVE (1944) and BAKER (1947) compiled a list of differential characters of both species. The most conspicuous differences between *M. album* and *M. dioicum* are found in flower characters. The white flowers of *M. album* open at night and have a greater diameter than the reddish purple flowers of *M. dioicum* which are open at daytime. The leaves of *M. album* are elliptical-lanceolate, the stem leaves of *M. dioicum* are ovate. An important difference is the overwintering system. *M. album* has a small rosette and large roots, whereas *M. dioicum* has a much bigger rosette but small roots. The leaves of *M. dioicum* are of a darker green than the leaves of *M. album*. The two species have different habitats. *M. album* is mainly found along roadsides and in open places that are or where recently cultivated, *M. dioicum* prefers a wooded environment.

We started our experiments with material collected in their original habitats. In the Netherlands seed from *M. album* was collected from a large population on a beet-field near Utrecht (R 255), along a roadside near Naarden (R 259), on an irregularly cultivated open space in the wood near Hilversum (R 460), near the dunes on the isles of Voorne (R 287) and "de Beer" (R 490), on beet-fields on Overflakkee (R 289) and on dikes on Goeree (R 290). In France, seed samples were gathered along the road Montpellier–Palavas (R 327), in Italy near La Sarre, south of Aosta at a height of 600 m. above sealevel (R 470). *M. dioicum* was collected in the Netherlands near Epe (Zuid Limburg) under hedges and on the border of wood (R 401, R 437), in bushes near Warmond (R 436), Naarden (R 260–262), Dedemsvaart (R 330), and in the dunes near Leiden (R 488). In France, *M. dioicum* was collected near Bonneval on a "Hochstauden" meadow

1900 m. above sealevel (R 471), along the brook Lenta (R 472, 1850 m.), and under shrubs (R 473, 1950 m.). In Italy, seed of *M. dioicum* was sampled near Cogne on a "Hochstauden" meadow (R 465, 1650 m.). In Switzerland, seeds were taken from plants growing in the surroundings of Klosters (R 489, 2800 m.). Professor WESTERGAARD kindly supplied us with seeds from tetraploid ♂ *M. album* with the mutated Y^1 chromosome (R 319).

Until 1961 we had no greenhouse at our disposal. The seeds were sown in sowing pans which were kept in frames in the experimental garden. After germination the seedlings were transplanted into pots of pressed soil which were subsequently planted out into the garden at the proper time. Half of the material was sown in September and after transplantation kept in frames during the winter. The seedlings were planted out into the garden in April of the following year. The second half of the material was sown in March and planted out into the field about May–June. The first group flowered about June–July, the second group during the months of July and August. This spreading of flowering facilitated the ordering of our experiments.

When in 1961 glasshouses became available, all the material was sown in a hothouse, the first group in February, the second in March. This way we got the same spreading in flowering time as before. In August just after harvesting, part of the material was sown, and planted out in October in a hothouse and an unheated glasshouse. In the former, flowering started at the end of January, in the latter in March–April. The seed harvested in the hothouse was sown immediately, and frequently a second flowering generation was obtained from it in the same year.

In our experimental garden *M. album* and *M. dioicum* were grown under identical conditions on a rather heavy clay soil in the open field. Both species grew very well.

Before crossing, male and female flowerbuds were isolated with a paperbag. When the flowers were open, the male flowers were removed from the plant and the crosses were performed by brushing the pistils with the stamens of the male flowers. After seedsetting was evident, the paperbags were removed. Each capsule was marked with a coloured thread.

Material originating from outside the experimental garden is registered with R followed by a number. Seedsamples from crosses performed in our garden are denoted by a letter, a different letter for each year, preceded by a number.

Results

SEX-LINKED INHERITANCE

A gene for narrow leaves

The recessive gene *angustifolia* (BAUR 1911, SHULL 1914) causes strong inhibition of growth in width of the leaves. WINGE (1927) demonstrated that the mutated X-chromosome with the gene *an-*

gustifolia caused sterility in the pollen containing this chromosome, no female plants with the angustifolia phenotype were found.

In our material of *M. album*, we also obtained a sex-linked gene which is responsible for the development of very narrow leaves, especially in the rosette stage. Our gene also affects the shape of the petals.

We first found three rosettes with narrow leaves in a family (76 J) selected for narrow petals. One of these plants, a male plant, came into flower. Obviously the pollen of this plant was rather sterile, since the first year pollinations did not result in seedsetting. The second year we obtained some progeny, all plants, males and females, were normal. In the next generation there were male plants with narrow leaves and male and female plants with normal leaves (Table 1). In other crosses also female plants with narrow leaves were obtained. Although phenotypic resemblance with the *angustifolia* type is striking, our gene for narrow leaves (*f*) is not necessarily identical with the gene (*b*) reported by BAUR. His gene is lethal in pollen, since only male plants have been discovered by WINGE in the progeny of an X_b Y plant. In our case males and females occur, though the fertility of the pollen both with X_f and with Y had considerably decreased.

TABLE 1

CROSSES DEMONSTRATING X-LINKED INHERITANCE OF A RECESSIVE GENE CAUSING NARROW LEAVES (n.l.) IN *M. album*

Parents	Offspring			
	Observed norm. n.l.	Expected norm. n.l.	Observed norm. n.l.	Expected norm. n.l.
XX (norm.) × X_fY (n.l.)	146 —	146 —	83 —	83 —
X_fX (norm.) × XY (norm.)	222 —	222 —	91 97	94 94
X_fX (norm.) × X_fY (n.l.)	23 29	26 26	15 11	13 13
$X_f X_f$ (n.l.) × X_fY (n.l.)	— 118	— 118	— 80	— 80

The gene abnormal

WINGE described the recessive gene *abnormal*. In our material we very often came across a mutant which is probably identical with *abnormal*. The expressivity of the gene is variable. Often the plant starts flowering with rather normal flowers, but in the later flowers the petals diminish in size. Notably in male plants this leads to the development of an inflorescence consisting of closed flowerbuds only. With the reduction in petal size the stamina become smaller and sterile. This gene seems to be very common because it frequently occurred as a disturbing inbreeding effect both in *M. album* and *M. dioicum*.

Certation

In *Melandrium*, deviations from the expected 1/1 ratio, female to male are due to different velocities in growth of the pollen tubes from gametes with an X- and gametes with a Y-chromosome, as was already demonstrated by CORRENS (1917, etc.). HARTSHORNE (1963) found no certation.

TABLE 2

SEX RATIOS OF VARIOUS CROSSES IN *Melandrium*

Type of cross		Number of families counted	Total number of plants	sex ratio (♀/♂)	
				mean	extremes
M. album	× *M. album*	19	1871	1.5	0.9 — 2.9
M. dioicum	× *M. dioicum*	11	1129	1.6	0.5 — 2.8
M. album	× *M. dioicum*	6	913	2.3	1.6 — 3.0
M. dioicum	× *M. album*	1	119	1.2	
M. (a. × d.)	× *M. (a. × d.)*	5	395	2.6	1.2 — 4.9
M. album	× *M. (a. × d.)*	1	92	1.2	
M. (a. × d.)	× *M. album*	1	48	0.8	
M. d. × (d. × (d. × (d. × (d. × a.))))		5	435	2.6	1.1 — 3.3

We did find a certation effect in *M. album*, *M. dioicum* and in various crosses where different combinations of sex chromosomes of both species were involved (Table 2). The variation in sex ratios of the different families will partly be due to the uncontrolled conditions during pollination, such as temperature, humidity, etc. Moreover, the length of the style and the quality of the pollen might be important factors. It is clear, however, that male gametes with an X-chromosome succeed more often in fertilizing a female gamete than do gametes with a Y-chromosome.

Genetic differences between X-chromosomes or between Y-chromosomes may also contribute to the variation in sex ratios. That an

X-chromosome can be responsible for an extreme ratio is demonstrated by a number of crosses (Table 3). From these crosses, we obtained female offspring only, with two exceptions (113 K, 35 L). Among their ancestors the male partners of these crosses have in common the female *M. dioicum* plant R 436II 18. The Y-chromosomes in these males are from four different origins (Table 3). It is highly unlikely that these Y-chromosomes, which behave normally in other crosses, are responsible for the extreme certation effect. Neither can the female partners of the crosses of table 3 be responsible, because they are also from very different origins, namely *M. album*, *M. dioicum*, *M.d.* × *M.a.*, and a plant (5R) which is a special flower colour type selected from a F2 generation of *M.a.* × *M.d.* These data show that the X-chromosome originating from R 486II 18 must cause this remarkable result. The fairly large number of male descendants in 35 L may be explained by the low fertility of the male parent of this cross, which fertility will have the same effect as pollination has with few pollen, which also works against certation.

TABLE 3

CROSSES SHOWING AN ABNORMAL SEX RATIO CAUSED BY A SPECIAL X-CHROMOSOME PRESENT IN THE MALE PARENTS, ORIGINATING FROM THE ♀ *M. dioicum* PLANT R 436 II 18.

Family number and cross	Origin of Y-chromosome of ♂ parent	Offspring ♀	♂
21 L *M. dioicum* × *M. dioicum*	R 436	40	—
35 L *M. (d.* × *a.)* × *M. (d.* × *a.)*	R 327	67	22
75 L *M. album* × *M. (d.* × *a.)*	,,	185	—
76 L *M. album* × *M. (d.* × *a.)*	,,	190	—
108 L 5 R *) × *M. (d.* × *a.)*	R 289	36	—
109 L *M. dioicum* × *M. (d.* × *a.)*	,,	45	—
110 L *M. dioicum* × *M. (d.* × *a.)*	,,	38	—
111 L *M. album* × *M. (d.* × *a.)*	,,	43	—
112 K *M. (d.* × *a.)* × *M. (d.* × *a.)*	R 255	92	—
113 K *M. (d.* × *a.)* × *M. (d.* × *a.)*	,,	81	1
175 K *M. (d.* × *a.)* × *M. (d.* × *a.)*	,,	116	—

*) R is a special flower colour type originally selected out of an F_2 generation of *M. album* × *M. dioicum*.

SEX-INFLUENCED INHERITANCE

Petals

Many measurements were carried out in *M. album* to investigate the variation in size and shape of the petal laminae. The length/width ratio of the petal-lobes was used as an indication of shape. The length and width of the lamina of one petal half pro plant were measured to the nearest tenth of a mm (Fig. 1). The petal halves when collected were stuck on transparent tape. Subsequently measurements were made with a Leitz binocular loupe (10 ×) leading the tape with petal halves over graph paper. Petal halves were taken because whole petals, especially broad ones could not be stuck on tape without wrinkling, which made exact measurements impossible.

When from a family at least 20 measurements pro sex were carried out, as a rule no big differences between the means of different samples of 20 measurements of the same family were found. The variation between the means is of course dependent on the heterogeneity of the material. The differences between the mean l/w ratios of different samples of one family never exceeded 5% of the largest mean l/w value observed in that family.

The significance of the differences occurring between the sexes was tested by a modified Wilcoxon test, P > 0.05 difference between the sexes is regarded as not significant.

TABLE 4

THE *l/w* RATIO ($\Sigma l/\Sigma w$), AND MEAN LENGTH AND WIDTH (IN mm) OF PETAL HALVES OF UNSELECTED *M. album* POPULATIONS

Populations		N	$\Sigma l/\Sigma w$	P	$\bar{l}$	$\bar{w}$
R 255	♀	63	2.13	0.082	12.36	5.79
	♂	72	2.05		13.38	6.53
R 327	♀	27	2.14	0.074	10.30	4.81
	♂	43	2.04		10.95	5.36
R 470	♀	36	2.46	0.032	11.61	4.71
	♂	74	2.32		12.91	5.56
R 490	♀	21	2.06	0.006	9.73	4.72
	♂	48	1.82		10.04	5.50
R 460	♀	23	1.95	0.002	13.10	6.71
	♂	22	1.70		12.89	7.59

TABLE 5

THE l/w RATIO ($\Sigma l/\Sigma w$), AND MEAN LENGTH AND WIDTH (IN mm) OF PETAL HALVES of *M. album* FAMILIES SELECTED FOR NARROW PETALS

Generations *)		N	$\Sigma l/\Sigma w$	P	$\bar{\imath}$	$\bar{w}$
R 255	♀	63	2.13	0.082	12.36	5.79
	♂	72	2.05		13.38	6.53
E (4)	♀	293	2.99	< 0.001	13.03	4.35
	♂	301	2.58		13.30	5.15
F (2)	♀	51	3.19	< 0.001	13.80	4.33
	♂	55	2.61		13.45	5.16
G (3)	♀	80	3.08	< 0.001	12.93	4.20
	♂	70	2.85		13.63	4.79
H (6)	♀	225	2.95	0.004	12.35	4.19
	♂	164	2.72		13.16	4.84
J (6)	♀	246	3.11	< 0.001	12.19	3.92
	♂	249	2.83		12.22	4.31
K (5)	♀	211	3.72	< 0.001	12.83	3.45
	♂	162	3.41		12.72	3.73

*) Selection started in the population R 255 and was continued during 6 subsequent generations (E–K). Number of families raised pro generation between brackets.

TABLE 6

THE l/w RATIO ($\Sigma l/\Sigma w$), AND MEAN LENGTH AND WIDTH (IN mm) OF PETAL HALVES OF *M. album* FAMILIES SELECTED FOR BROAD PETALS

Generations *)		N	$\Sigma l/\Sigma w$	P	$\bar{\imath}$	$\bar{w}$
R 255	♀	63	2.13	0.082	12.36	5.79
	♂	72	2.05		13.38	6.53
E (2)	♀	64	1.79	0.061	15.36	8.60
	♂	71	1.69		15.01	8.89
F (8)	♀	207	1.75	0.065	15.99	9.11
	♂	238	1.67		14.66	8.77
G (7)	♀	212	1.49	0.534	13.68	9.19
	♂	181	1.46		13.54	9.29
H (12)	♀	335	1.48	0.439	13.54	9.13
	♂	363	1.53		13.09	8.56
J (8)	♀	354	1.43	0.671	14.90	10.39
	♂	360	1.38		14.50	10.51
K (4)	♀	98	1.30	0.545	15.26	11.76
	♂	128	1.28		14.40	11.27

*) Selection started in the population R 255 and was continued during 6 generations (E–K). Number of families raised pro generation between brackets.

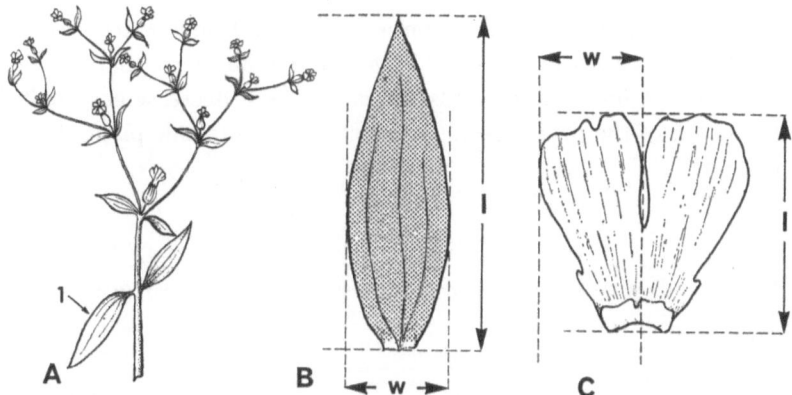

Fig. 1. Leaves on the first node beneath the inflorescence on the main stem were measured (A–1). Length and width of leaves and petal halves were determined as shown (B and C).

In table 4 data are given of measurements of some populations of *M. album*, raised in our garden from seed collected on various habitats. In this unselected material the female plants have narrower petals (higher l/w value) than the male plants of the same family.

A two-way selection program was started in population R 255. In this population individuals with the narrowest petals were crossed with each other. The petals of the resulting families were measured. In the family which on the average had the narrowest petals, again the individuals with the narrowest petals were selected for further crosses. A similar method was used for families with broad petals. These selections were done for 6 generations.

Selection for narrow petals seems to increase the difference between male and female plants (Table 5). After selection for broad petals the difference disappears (Table 6).

When we compare families with broad petals (Table 6) and families with narrow petals (Table 5) it appears that narrow petals are at the same time slightly shorter than broad ones. In families with broad petals, the female petals are larger than the male petals, but have the same l/w ratios. In families with narrow petals the petals of the female flowers are smaller than those of the male flowers, but have a higher l/w ratio than the latter, due to the stronger decrease in width. An accumulation of genes causing narrow petals thus seems to have a stronger effect in female plants than in male plants.

The data given pro generation in table 5 and table 6 are presented pro family in a graph (Fig. 2). This graph too illustrates that the effect of the genes for narrow petals, which causes a decrease especially in the width of the petals, seems to be stronger in female plants than in male ones.

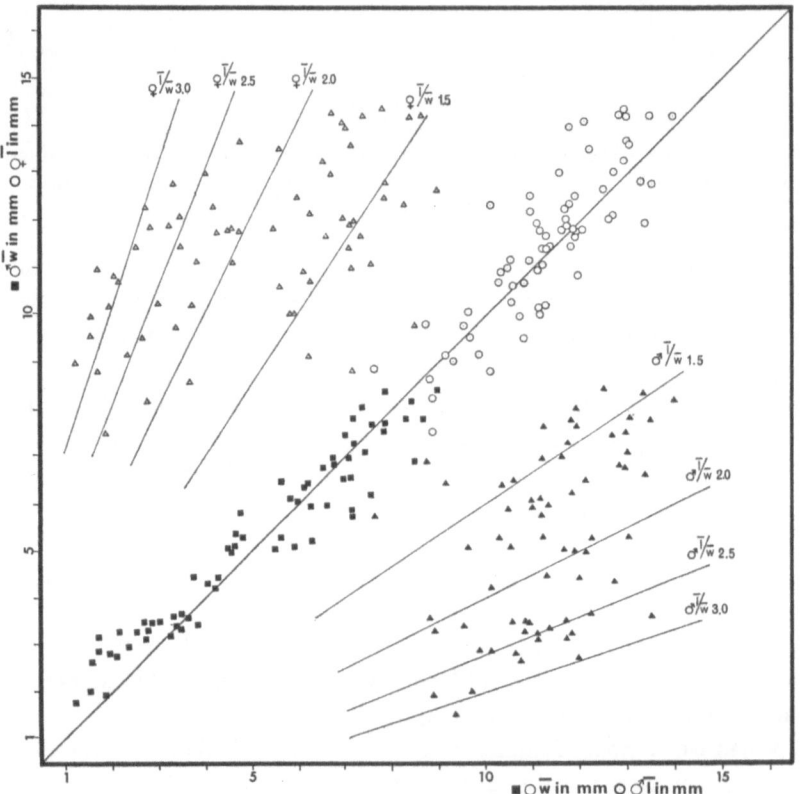

Fig. 2. The ♀ l̄, ♀ w̄, ♂ l̄ and ♂ w̄ of petal halves of a number of *M. album* families are used as coordinates in various combinations.

  ○ Abscissa: ♂ l̄ (in mm); ordinate: ♀ l̄ (in mm).
  ■ Abscissa: ♀ w̄ (in mm); ordinate: ♂ w̄ (in mm).
  ▲ Abscissa: ♂ l̄ (in mm); ordinate: ♂ w̄ (in mm).
  △ Abscissa: ♀ w̄ (in mm); ordinate: ♀ l̄ (in mm).

Notably in families with a small w̄, the ♂ w̄ is larger than the ♀ w̄ of the same family.

Leaves

Size and shape of leaves were determined in *M. album* and *M. dioicum*. Measurements were performed in both species on unselected material and in *M. album* also on families selected for narrow leaves and on other families selected for broad leaves. The same selecting method was used as in selection for petal-shape.

Pro plant the length and width was determined in mm, of one stem leaf on the first node below the inflorescence on the main stem (Fig. 1A and B).

With regard to shape, the leaves of *M. album* display the same tendencies as the petals. In unselected material from various origins, the $\Sigma l/\Sigma w$ ratio is higher in female plants than in the male (Table 7). After selection for narrow leaves (Table 8) the difference becomes more distinct. Selection for broad leaves comprises two generations only (Table 9). This is probably the reason why the difference between the sexes still exists. The populations R 470 and R 490 (Table 7) have,

TABLE 7

THE *l/w* RATIO ($\Sigma l/\Sigma w$), MEAN LENGTH AND WIDTH (IN mm) OF LEAVES OF UN-
SELECTED *M. album* FAMILIES FROM VARIOUS ORIGINS

Original populations *)		N	$\Sigma l/\Sigma w$	P	$\bar{l}$	$\bar{w}$
R 255 (14)	♀	851	4.51	< 0.001	69.96	15.51
	♂	672	3.88		58.36	15.05
R 259 (6)	♀	317	4.24	< 0.001	75.60	17.81
	♂	290	3.43		54.00	15.73
R 289 (2)	♀	91	5.23	< 0.001	64.89	12.40
	♂	133	4.25		58.44	13.75
R 290 (2)	♀	101	3.39	< 0.001	68.68	20.91
	♂	81	2.89		60.20	22.37
R 327 (5)	♀	325	3.55	0.021	56.93	16.04
	♂	166	3.25		42.68	13.15
R 460 (2)	♀	213	4.37	< 0.001	56.60	12.94
	♂	261	3.57		43.76	12.25
R 470 (4)	♀	174	2.82	0.545	55.75	19.80
	♂	238	2.84		50.06	17.64
R 490 (0)	♀	45	2.13	0.675	60.58	28.44
	♂	42	2.14		53.95	25.17

*) Number of families between brackets.

TABLE 8

THE l/w RATIO ($\Sigma l/\Sigma w$), MEAN LENGTH AND WIDTH (IN mm) OF LEAVES OF $M.$ *album* FAMILIES SELECTED FOR NARROW LEAVES DURING 5 (F–K) AND 3 (H–K) GENERATIONS

Original populations	Generations *)		N	$\Sigma l/\Sigma w$	P	l	$\overline{w}$
R 255	F (4)	♀	156	5.18	0.005	74.34	14.35
		♂	149	4.59		62.46	13.60
	G (3)	♀	90	4.90	0.033	†)	
		♂	121	4.72			
	H (8)	♀	277	5.18	0.046	77.62	14.99
		♂	271	5.06		65.00	12.84
	J (5)	♀	229	7.19	< 0.001	63.89	8.89
		♂	274	6.35		55.81	8.78
	K (2)	♀	56	7.31	< 0.001	67.38	9.21
		♂	59	6.25		53.32	8.53
R 289	H (2)	♀	150	5.34	< 0.001	61.96	11.59
		♂	75	3.72		48.07	12.92
	J (3)	♀	134	6.78	< 0.001	72.89	10.75
		♂	100	4.55		60.47	13.29
	K (3)	♀	151	6.06	< 0.001	59.39	9.79
		♂	101	5.32		52.73	9.91

*) Number of families pro generation between brackets.
†) Data are lost.

TABLE 9

THE l/w RATIO ($\Sigma l/\Sigma w$), MEAN LENGTH AND WIDTH (IN mm) OF LEAVES OF $M.$ *album* FAMILIES SELECTED OUT OF R 255 AND R 327 FOR BROAD LEAVES DURING 2 GENERATIONS

Original populations	Generations *)		N	$\Sigma l/\Sigma w$	P	$\overline{l}$	$\overline{w}$
R 255	J (2)	♀	108	3.48	0.017	83.60	24.02
		♂	129	3.11		76.55	24.62
	K (4)	♀	88	2.99	0.044	88.95	29.80
		♂	111	2.83		76.31	26.93
R 327	J (3)	♀	229	3.38	0.018	59.18	17.52
		♂	112	3.11		51.65	16.63
	K (4)	♀	168	2.92	0.027	54.30	18.58
		♂	156	2.72		45.04	16.53

*) Number of families pro generation between brackets.

TABLE 10

THE l/w RATIO ($\Sigma l/\Sigma w$), MEAN LENGTH AND WIDTH (IN mm) OF LEAVES OF
M. dioicum POPULATIONS

Populations		N	$\Sigma l/\Sigma w$	P	$\bar{l}$	$\bar{w}$
R 191	♀	193	1.65	0.196	60.63	36.69
	♂	102	1.59		48.64	30.56
R 401	♀	105	2.31	0.072	61.62	26.70
	♂	119	2.52		51.24	20.37
R 436	♀	113	2.01	0.535	61.28	30.51
	♂	61	2.01		54.89	27.25
R 437	♀	120	2.12	0.603	55.77	26.27
	♂	146	2.13		45.92	21.57
R 488	♀	30	1.99	0.590	51.27	25.77
	♂	49	2.04		43.74	21.41
M.d. Zuid	♀	272	1.88	0.404	55.65	29.48
Limburg *)	♂	283	1.83		49.22	26.96

*) Measured on the original habitat.

TABLE 11

THE l/w RATIOS ($\Sigma l/\Sigma w$) OF LEAVES IN DIPLOID *M. album* FAMILIES WITH FEMALE
PLANTS AND HERMAPHRODITES

Family		N	$\Sigma l/\Sigma w$	P
195 H (197 F)	♀ (♀)	47 (31)	4.65 (4.66) *)	0.030 (0.002)
	⚥ (♂)	39 (32)	4.30 (4.27)	
196 H (104 G)	♀ (♀)	45 (77)	3.79 (3.86)	< 0.001 (0.011)
	⚥ (♂)	22 (84)	3.03 (3.24)	
195 J (198 F)	♀ (♀)	37 (45)	4.75 (4.61)	< 0.001 (0.002)
	⚥ (♂)	58 (45)	4.01 (4.14)	
196 J (90 H)	♀ (♀)	49 (50)	4.21 (4.24)	< 0.001 (< 0.001)
	⚥ (♂)	51 (35)	3.64 (3.44)	
200 J (199 F)	♀ (♀)	60 (49)	5.35 (5.47)	< 0.001 (< 0.001)
	⚥ (♂)	58 (39)	4.68 (5.08)	

*) Data between brackets show comparable ratios in families with female and
male plants.

TABLE 12

THE l/w RATIOS $(\Sigma l/\Sigma w)$ OF LEAVES IN TETRAPLOID
M. *album* FAMILIES WITH FEMALE PLANTS AND HERMA-
PHRODITES AND TWO TETRAPLOID FAMILIES WITH FEMALE
AND MALE PLANTS

Family		N	$\Sigma l/\Sigma w$	P
R 319	♀	31	2.93	< 0.001
	⚥	49	2.24	
37 G	♀	20	2.63	< 0.001
	⚥	32	2.13	
138 H	♀	33	2.70	0.0015
	⚥	63	2.43	
139 H	♀	51	2.64	0.002
	⚥	83	2.35	
142 H	♀	16	3.08	< 0.001
	⚥	70	2.54	
146 H	♀	37	2.61	< 0.001
	⚥	82	2.17	
27 H	♀	48	2.82	< 0.001
	♂	62	2.54	
199 H	♀	57	3.55	
	♂	57	3.31	0.030

without selection, broad leaves and no differences between the $\Sigma l/\Sigma w$ ratios of the sexes were observed. Also *M. dioicum* with its broad leaves displays no differences between the $\Sigma l/\Sigma w$ ratios of female plants and male ones (Table 10).

The leaves of female plants of *M. album* and *M. dioicum* are on the whole larger than the leaves of male plants.

In a number of *M. album* families with female plants and herma-phrodites, the latter caused by an abnormal Y-chromosome called Y^a (NIGTEVECHT, 1966), l/w ratios of leaves were determined (Table 11). The differences between the l/w ratios of females and hermaphrodites were equal to the differences between the l/w ratios of female plants and male plants in normal families with comparable leaf shapes. In tetraploid families, female plants and hermaphrodites also showed a significant difference in l/w ratios (Table 12). In these families an abnormal Y-chromosome (Y^1) found by WESTERGAARD (1946) caused the hermaphroditism.

Anthocyanin formation

The green parts of plants of *M. album* and of *M. dioicum* are often stained reddish brown by anthocyanin. We noticed that especially *M. album* is very variable in this respect. It is known that in general, anthocyanin formation is strongly dependent on light, so that plants growing on shaded places usually have far less anthocyanin than individuals on sunny sites. However, since in the experimental garden all plants were grown in the open, the differences noted cannot have been due to such an effect.

A classification of anthocyanin content in the calyx and in the stem was made as follows:

Anthocyanin in the calyx.

0. No anthocyanin.
1. Main veins with anthocyanin only.
2. Main and side veins with anthocyanin.
3. Also between the side veins some anthocyanin.
4. Between the side veins very much anthocyanin.

Anthocyanin in the stem.

0. No anthocyanin.
1. Anthocyanin in the very first node of the stem only.
2. Anthocyanin in nearly all nodes of the stem.
3. Anthocyanin in nearly all nodes of the stem and some anthocyanin in the internodes.
4. Internodes with very much anthocyanin.

An analysis was made of a number of randomly chosen *M. album* families (Table 13). There appeared to be a difference in anthocyanin content between the sexes, male plants having more anthocyanin in calyx and stem than female plants.

As the genes for anthocyanin formation have constantly a stronger effect in male plants than in the female irrespective whether the family as a whole displays much or little anthocyanin in the green parts, we may consider these genes to follow a sex-influenced inheritance.

Also anthocyanin formation in the petals appears to be sex-influenced. The petals of *M. album* are as a rule without any anthocyanin. When some plants in a population of *M. album* resulting from introgressive hybridisation with *M. dioicum* have very faint-red coloured

TABLE 13

PLANTS OF DIPLOID (2n) AND TETRAPLOID (4n) FAMILIES OF *M. album* CLASSIFIED
ACCORDING TO ANTHOCYANIN CONTENT IN CALYX AND STEM. INTENSITY CLASSES
0–4, SEE TEXT

Family		N	Numbers of plants classified to anthocyanin content in calyx					Numbers of plants classified to anthocyanin content in stem				
			0	1	2	3	4	0	1	2	3	4
17 E	♀	33	17	5	10	1	—					
2n	♂	38	—	—	11	27	—					
30 E	♀	144	38	40	61	5	—					
2n	♂	106	6	3	26	71	—					
209 G	♀	56	—	9	18	19	10	—	2	33	21	—
2n	♂	50	—	—	—	—	50	—	—	10	37	3
210 G	♀	8	2	3	2	1	—	—	—	5	3	—
2n	♂	36	—	1	13	7	15	—	1	13	20	2
171 H	♀	58	—	—	26	29	3	—	—	54	4	—
2n	♂	71	—	—	4	18	49	—	—	61	10	—
173 H	♀	48	—	5	29	9	5	—	—	46	2	—
2n	♂	63	—	—	5	35	23	—	—	60	3	—
138 H	♀	33	1	—	12	20	—	—	1	28	4	—
4n	☿	63	—	1	3	29	30	—	—	50	13	—
142 H	♀	15	—	—	4	11	—	—	—	14	1	—
4n	☿	69	—	—	1	54	14	—	—	57	12	—
146 H	♀	37	—	1	17	18	1	—	—	29	8	—
4n	☿	82	—	1	17	63	1	—	—	53	28	1

petals, these plants are mainly male plants. Possible differences in anthocyanin formation between male and female petals of *M. dioicum* cannot be demonstrated by comparing the intensities of the deep reddish purple coloured petals. In the lighter coloured offspring of *M. album* × *M. dioicum* crosses, the male petals appear to be somewhat more vividly coloured than the female petals.

We found another way of comparing anthocyanin formation in female and male petals in *Melandrium*. The pH of the cell sap of the petals appeared to fluctuate with the colour intensity, possibly due to the fact that the anthocyanins present in the material studied were acylated. The more anthocyanin is present the lower the pH. In crosses not mentioned here no independent gene for pH determination could be detected.

Measurements were performed on the pH of an extract of the petals of *M. album, M. dioicum, M. album* × *M. dioicum* and *M. dioicum* × *M. album* (Table 14). Pro family the pH of male petals was determined on a sample consisting of petals from all male plants. The pH of female petals was determined analogously. Measurements were made on fresh ground petals in doubly distilled water. A solution was made equivalent to 2 grams of petals in 30 ml distilled water.

In *M. album* no consistent difference between female and male plants was found. The pH of the extract was about 6.0 in both sexes. The same pH was found in the extract of the leaves of *M. album* and of *M. dioicum*. When anthocyanin was present in the petals the pH

TABLE 14

CORRELATION BETWEEN COLOUR INTENSITY (RANGING FROM
0 TO 5) AND pH OF PETAL EXTRACTS IN *Melandrium*

Family	Colour intensity of petals	pH		Number of plants	
		♀	♂	♀	♂
M. album					
47 J	0	6.0	5.9	20	46
131 J	0	6.0	5.9	39	36
175 J	0	5.9	6.1	35	32
269 J	0	5.8	6.0	15	87
270 J	0	6.0	6.0	22	64
M. dioicum					
222 J	5	5.2	5.1	23	20
230 J	4	5.4	5.3	10	59
M. a. × *M. d.*					
224 J	3	5.7	5.5	49	49
226 J	3	5.6	5.5	46	50
227 J	3	5.6	5.4	51	52
M. d. × *M. a.*					
212 J	3	5.5	5.3	44	53
214 J	2–3	5.8	5.5	59	58
215 J	2–3	5.8	5.5	21	14

Without anthocyanin a pH of approximately 6.0 was observed. Families with anthocyanin in the petals show a higher pH of the cell sap in the petals of the female plants than of the male plants.

of the petals was always lower than the pH of the leaves and the pH in male petals was always lower than the pH in female petals of the same family.

These data support our presumption that the genes that determine the amount of anthocyanin in the petals behave like sex-influenced genes.

Glandular hairs

In most cases *M. album* plants have glandular hairs. The density of these hairs is highest on the top of the plants. Downward a decrease in number is seen. With a decrease in number the length of the hairs and the diameter of the top cells become at the same time smaller. When no glandular hairs are found on the calyces, the whole plant will lack these hairs.

When all the plants of a family have many glandular hairs it is very difficult to distinguish different categories. In families where the glandular hairs are less abundant an analysis can be made. In R 490, R 470 and the families 55 K, 76 K and 77 K raised from R 470, on the whole, less glandular hairs were observed, than in many other families of *M. album*. They had many normal hairs instead. In these families it was possible to classify the plants according to the relative amount of glandular hairs on the calyx. The data (Table 15) show a difference between both sexes. As a rule female plants have fewer glandular hairs than male plants of the same family. This is demonstrated by

TABLE 15

PRESENCE OF GLANDULAR HAIRS ON *M. album* PLANTS OF R 470, R 490 AND OF THREE FAMILIES RAISED FROM R 470

	Percentage of plants with glandular hairs		Mean relative amount of glandular hairs on calyces of plants with these hairs *)	
	♀	♂	♀	♂
R 490	18	83	1.5	2.0
R 470	18	75	1.7	1.9
55 K	82	100	1.8	2.5
76 K	40	89	1.5	2.1
77 K	6	60	1.0	1.7

*) Estimation 0, no glandular hairs, to 3, many glandular hairs.

the different percentages of female plants and male plants with glandular hairs and also by the difference in relative amounts of glandular hairs on female and male plants with glandular hairs.

The genes for glandular hairs obviously behave differently in the two sexes. In female plants these genes have a lower expressivity than in male plants. Therefore, when the plants have a genotype for few glandular hairs the character might show up in the male plants and not in the female ones.

TABLE 16

CROSSES BETWEEN *M. album* PLANTS WITHOUT GLANDULAR
HAIRS SELECTED OUT OF 77 K (TABLE 15)

Family	Glandular hairs *)									
	♀					♂				
411 M	+	−	−	−	−	1–2	1	1	+	+
412 M	−	−	−	−	−	1	+	+	−	−
413 M	−	−	−	−	−	1–2	1	−	−	−
414 M	−	−	−	−	−	1	1	1	+	+
415 M	+	−	−	−	−	+	+	+	−	−
415 MA	−	−	−	−	−	1	1	1	1	+
416 M	−	−	−	−	−	1	+	+	+	−
417 M	+	−	−	−	−	1	1	+	−	−
418 M	−	−	−	−	−	1	1	1	+	+
419 MA	−	−	−	−	−	+	+	+	+	−

*) Glandular hairs determined on calyces of 5 female and 5 male plants pro family. −, no glandular hairs; +, sometimes a glandular hair is found; 1, one row of glandular hairs on the main veins; 2, more than one row of glandular hairs on the main veins.

A number of special crosses between plants without glandular hairs (Table 16) displays the same difference in penetrance. Out of 10 families 5 individuals pro sex were observed very carefully. Among these plants 47 females were found without glandular hairs and 3 with a few, where 11 male plants had none and 39 some glandular hairs. Apparently the genes for glandular hairs which had no penetrance in the female parents of these families did have some effert in most male offspring.

These genes can therefore be considered as sex-influenced genes.

Discussion

Sex-linked genes are genetically linked to the mechanism of sex determination. Sex-influenced genes might be regarded as physiologically linked to the same mechanism.

Sex-linked genes are by definition genes that are situated in the sex chromosomes, but take no part in sex determination. We described the gene in the X-chromosome causing narrow leaves (*f*), as a sex-linked gene. However, *f* influences sex expression, since notably in male plants the fertility has decreased markedly. In fact the same holds true for the gene *abnormal*. Therefore both genes might be regarded as sex-determining factors. In the differential part of one or both sex chromosomes, genes must be present that influence the growth of the pollen tubes, thus giving rise to the certation effect. An extreme certation effect, caused by a special X-chromosome, was demonstrated by crosses that yielded female offspring only (Table 3). As there is at the moment no proof that the genes causing certation are sex-determining genes they must be called sex-linked genes.

Sex-influenced genes play no part in sex determination, but their effect is influenced by the sex-determining genes, in such a way that when the inner milieu is suitable for the development of a gynoecium only, the phenotype brought about by the sex-influenced genes is not the same as when the sex-determining mechanism leads to the development of the male organs only.

Sex-influenced inheritance leads to sex dimorphism. This dimorphism is not restricted to situations where two different genotypes are present, as in dioecious species. The monoecious *Zea mays* has male and female inflorescenses, which differ clearly from each other.

The dioecious orchid *Catasetum barbatum* produces every now and then a monoecious plant (GOEBEL, 1913). The female and male flowers differ so profoundly that before the monoecious individuals were found male plants and female were described as belonging to different genera.

The influence of the sex-determining mechanism on the sex influenced genes often extends beyond the flowers. Female plants of the dioecious species *Rubus chamaemorus* have five-lobed leaves on the vegetative shoots, and three-lobed leaves on the flowering shoots. Male plants have three-lobed leaves and single-lobed ones respectively.

Female plants in most dioecious species are larger than the male plants (GOEBEL, 1913). *Melandrium* is no exception to this rule. In *M. album* and in *M. dioicum* female plants are bigger and have also larger leaves than male plants. The presence of less anthocyanin in females than in males might also be an indication of a more vigorous growth of females, since anthocyanin often occurs where growth has been inhibited.

On the other hand, petals of male plants are larger than petals of female plants in unselected populations (Table 4).

This seems to be in contradiction with the greater over all size of the female plants. However, in dioecious species the male petals are often larger (GOEBEL, 1913). This might be explained by assuming that the same mechanism that in the female flowers inhibits the development of the stamina, suppresses to a lesser extent the develop-

TABLE 17

THE l/w RATIO $(\Sigma l/\Sigma w)$ OF LEAVES AND PETALS OF *M. album*
DEMONSTRATING INDEPENDENT INHERITANCE OF SHAPE OF
LEAVES AND SHAPE OF PETALS

	Leaves $\Sigma l/\Sigma w$		Petals $\Sigma l/\Sigma w$	
	♀	♂	♀	♂
No selection				
R 327	3.49	3.39	2.14	2.04
R 460	4.32	3.57	1.95	1.70
R 470	3.48	3.26	2.46	2.32
R 490	2.13	2.14	2.06	1.82
Selection for broad petals				
71 H	4.32	3.70	1.18	1.26
76 H	4.11	3.58	1.12	1.20
Selection for narrow petals				
270 F	4.16	3.64	3.49	2.54
114 G	4.29	3.68	3.30	2.90
Selection for broad petals and narrow leaves				
299 H	7.00	6.17	1.66	1.50
175 J	7.13	6.10	1.59	1.47
269 J	7.28	6.65	1.37	1.36
270 J	8.15	6.87	1.82	1.59

ment of the petals. In the discussion on sex determination (Part II) more will be said about this suppressing principle (NIGTEVECHT, 1966).

It is not clear by which physiological principle the sex determination mechanism influences the formation of glandular hairs and the shape of petals and leaves.

Obviously shape of leaves and shape of petals are not determined by the same genes (Table 17). Without selection distinct differences may exist between the l/w ratios of leaves of various populations; for instance R 460, ♀ l/w 4.32 and R 490, ♀ l/w 2.13. However, there is no difference in the l/w ratios of the petals of the same populations. By selection the discrepancy can be made even stronger. The families 71 H and 76 H are selected for broad petals, ♀ l/w 1.18 and 1.12 respectively. The families 270 F and 114 F are selected for narrow petals, ♀ l/w 3.49 and 3.30. These four families, however, do not differ in shape of leaves. A number of other families could be selected for narrow leaves and broad petals at the same time, 269 J ♀ l/w leaves 7.28, petals 1.37.

We assume therefore that different gene complexes are involved in petal shape and leaf shape determination. However, both gene complexes are influenced by the sex determination mechanism in the same direction.

The inhibition of growth in width caused by genes for narrow leaves and by other genes for narrow petals seemed to be stronger in female plants than in male ones. A series of measurements on leaves in families consisting of females and hermaphrodites demonstrated a similar difference between females and hermaphrodites (Tables 11, 12). Moreover, the calyces of the hermaphrodite plants contained clearly more anthocyanin than the calyces of the females in the respective families. So, with regard to these sex-influenced characters, males in *Melandrium* are more like hermaphrodites than females are.

These observations on sex-influenced characters might be of significance for an understanding of the physiological processes involved in sex determination, as the mechanism of sex determination expresses itself not only in the formation of the sex organs but also in the sex-influenced characters.

The results of our investigations on sex determination will be reported in part II of these "studies in dioecious *Melandrium*" (NIGTEVECHT, 1966).

REFERENCES

BAKER, H. G. (1947). *Melandrium album* and *M. dioicum* in the Biological Flora of the British Isles. *J. Ecol.* **35**: 271–292.

BAKER, H. G. (1948). The ecotypes of *Melandrium dioicum* (L. emend.) Coss. and Germ. *New Phytol.* **47**: 131–145.

BAKER, H. G. (1951). The inheritance of certain characters in crosses between *Melandrium dioicum* and *M. album*. *Genetica* **25**: 126–156.

BATESON, W. & E. R. SAUNDERS (1902). Experimental studies on the physiology of heredity. *Rep. to the Evol. Comm. of the Royal Soc.* **1**: 3–160.

BAUR, E. (1910). Untersuchungen über die Vererbung von Chromatophoren merkmalen bei *Melandrium, Antirrhinum* und *Aquilegia. Z.I.A.V.* **4**: 81–102.

BAUR, E. (1911). Einführung in die experimentelle Vererbungslehre. Borntraeger, Berlin.

BAUR, E. (1912). Ein Fall von geschlechtsbegrenzter Vererbung bei *Melandrium album. Z.I.A.V.* **8**: 335–336.

CORRENS, C. (1900). Ueber Levkoyenbastarde. Zur Kenntnis der Grenzen der Mendelschen Regeln. *Bot. Zentralbl.* **84**: 111–126.

CORRENS, C. (1903). Über die dominierende Merkmale der Bastarde. *Ber.Dtsch. Bot. Ges.* **21**: 133–147.

CORRENS, C. (1921). Zweite Fortsetzung der Versuche zur experimentellen Verschieben des Geschlechtsverhältnisses. *Sitzungsber. d. Preuss. Akad. d. Wiss. Phys.-Mat. Klasse* (1921): 330–354.

CORRENS, C. (1922). Alkohol und Zahlenverhältnis der Geschlechter bei einer getrennt geschlechtigen Pflanze (*Melandrium*). *Naturw.* **49**: 1049–1052.

CORRENS, C. (1924). Über den Einflusz des Alters der Keimzellen. I. *Sitzungsber. d. Preuss. Akad. d. Wiss.* (1924): 70–104.

CORRENS, C. (1926). Über Fragen der Geschlechtsbestimmung bei höheren Pflanzen. *Z.I.A.V.* **41**: 5–40.

GOEBEL, K. (1913). Organographie der Pflanzen. Gustav Fischer, Jena.

HARTSHORNE, J. W. (1963). The heterogametic sex in dioecious flowering plants. *Genetics today, Proc. XI Intern. Congr. Genetics* I: 232. Pergamon Press, Oxford.

LAWRENCE, C. W. (1963). Genetic studies on wild populations of *Melandrium*. I. Chromosome behaviour. *Heredity* **18**: 135–148.

LAWRENCE, C. W. (1963). Genetic studies on wild populations of *Melandrium*. II. Flowering time and plant weight. *Heredity* **18**: 149–163.

LAWRENCE, C. W. (1964). Genetic studies on wild populations of *Melandrium*. III. *Heredity* **19**: 1–19.

LÖVE, D. (1944). Cytogenetic studies on dioecious Melandrium. *Botan. Notiser* (1944): 125–214.

NIGTEVECHT, G. VAN (1966). Genetic studies in dioecious *Melandrium*. II. Sex determination in *M. album* and *M. dioicum*. *Genetica* **37**: 307–344.

SCHULZ, A. (1890). Beiträge zur Kenntnis der Bestäubungseinrichtungen und Geschlechtsverteilung bei den Pflanzen. II. *Bibl. Bot.* **17**: 182–196.

SHULL, G. H. (1910). Inheritance of sex in *Lychnis*. *Bot. Gaz.* **52**: 110–125.

SHULL, G. H. (1911). Reversible sex-mutants in *Lychnis dioica*. *Bot. Gaz.* **52**: 329–368.

SHULL, G. H. (1914). Sex-limited inheritance in *Lychnis dioica* L.. *Z.I.A.V.* **12**: 265–302.

STANFIELD, J. F. (1937). Certain physico-chemical aspects of sexual differentiation in *Lychnis dioica*. *Am. J. Bot.* **24**: 710–719.

STANFIELD, J. F. (1944). Chemical composition of roots and tops of dioecious *Lychnis* in vegetative and flowering phases of growth. *Plant Physiol.* **19**: 377–383.

TUTIN, T. G., V. H. HEYWOOD, N. A. BURGES, D. H. VALENTINE, S. M. WALTERS & D. A. WEBB (1964). Flora europaea. Cambridge Univ. Press. London.

VRIES, H. DE (1900). Das Spaltungsgesetz der Bastarde. *Ber. Dtsch. Bot. Ges.* **18**: 83–90.

VRIES, H. DE (1903). Die Mutationstheorie. Bd. **2**. Veit & Comp., Leipzig.

WESTERGAARD, M. (1946). Aberrant Y-chromosomes and sex expression in *Melandrium album*. *Hereditas* **32**: 419–443.

WINGE, Ø. (1917). The chromosomes. Their numbers and general importance. *Comptes rendus du Lab. Carlsberg* **13**: 131–175.

WINGE, Ø. (1927). On a Y-linked gene in *Melandrium*. *Hereditas* **9**: 274–284.

WINGE, Ø. (1931). X- and Y-linked inheritance in *Melandrium*. *Hereditas* **15**: 127–165.

WINGE, Ø. (1932). The nature of sex-chromosomes. *Proc. Sixth Intern. Congr. Genetics* **1**: 343–355.

Genetica (1966) **37**: 307–344

GENETIC STUDIES IN DIOECIOUS *MELANDRIUM*. II.

SEX DETERMINATION IN *MELANDRIUM ALBUM* AND *MELANDRIUM DIOICUM*

G. van Nigtevecht

Institute of Genetics, State University, Utrecht, The Netherlands
(Received April 1, 1966)

The process of sex determination leads to the formation of the sex organs. Formerly two different groups of sex-determining genes were distinguished: the sex-promoting genes causing the actual development of the sex organs, and sex-deciding genes, which decide whether female-promoting genes, or male-promoting genes, or both come into action, thus giving rise to female flowers, male and hermaphrodite ones respectively.

Observations on our own material and data from the literature lead us to the conclusion that no such sharp distinction could be made. In this respect especially the hypothesis of Heslop–Harrison (1957) on the role of auxin in sex determination is of interest. It is argued that the sex-determining genes together provoke in flower primordia a certain net auxin activity that determines the development of the sex organs. A genotype that realizes a relatively high net auxin activity will favour the development of pistils and will work against the formation of stamina. A genotype that causes a relatively low auxin activity has the opposite effect.

Changes of distinctly different sex-determining genes might have a similar effect on sex-expression. For example changes of genes in the Y-chromosome of an XY plant may cause the development of a pistil in the flowers of an XY plant and simultaneously work against the formation of stamina, as is demonstrated by some mutations in our *Melandrium* material. In one mutation cytologically no aberration of the Y-chromosome was observed. The Y-chromosome of the other mutations appeared to lack a part of their non-homologous arms. We observed the same phenotypical change after autosomal selection started in three *M. dioicum* populations.

A new feature of the distal part of the non-homologous arm of the Y-chromosome showed up in our hermaphrodite gerontogones. This part appeared to be indispensable for the functioning of both female and male gametes.

In the discussion the formation of unisexual flowers in monoecious species and in dioecious ones was regarded to be fully comparable. In both instances auxin was assumed to play a dominant part in the differentiation of the flowers. We discussed the possible role of auxin in the regulating mechanism of protein synthesis.

The origin of dioecism was explained in terms of building up an uneven distribution of + genes (increasing the net auxin activity) and − genes (decreasing the net auxin activity) in a non-homologous pair of chromosomes.

It is suggested that apomictic plants are like female plants characterized by a relatively high auxin activity in the flower primordia. We assumed therefore apomixis and dioecism to be by their origin related phenomena.

With regard to the discrepancy in occurrence of dioecism and polyploidy in the plant kingdom and in the animal kingdom, it is argued that in plant species polyploidisation interferes with the building up of a balanced dioecious system, a situation not met with in the animal kingdom, because of the unfavourable aspects of polyploidy in animals.

Introduction

The mechanism of sex determination in plants has been investigated very thoroughly in *Melandrium*.

Aberrant sex types, especially hermaphrodites played an important part in these investigations. The hermaphrodites thus far found in *Melandrium* belong to different types.

In one type an abnormality of the sex chromosome is the causing factor, as could be demonstrated genetically and sometimes cytogenetically. WINGE (1931) found hermaphrodites of this type in an F_2 generation from *M. album* and M. *dioicum*. WESTERGAARD (1946), detected abnormal Y-chromosomes in the progeny of triploid material of *M. album*, WARMKE (1946) in diploid XXY stocks. LÖVE (1942) found polygamous androhermaphrodites in *M. dioicum*, which she supposed to have a translocated X-chromosome. This X-chromosome should contain a part of the Y-chromosome with male promoting genes. ÅKERLUND (1927) described a sterile hermaphrodite of *M. dioicum*. On cytological grounds he assumed this hermaphrodite to have two X-chromosomes. This hypothesis could not be verified because no offspring was obtained.

A second type of hermaphrodites occurs when the chromosomal complement is out of balance. Hermaphrodites due to variation in autosome number have been found in crosses between tetraploid and triploid individuals (WESTERGAARD, 1948). WARMKE (1946) described some tetraploid hermaphrodites arising from an abnormal combination of otherwise normal sex chromosomes. In these tetraploids 4 X-chromosomes together with one Y-chromosome caused the development of stamina and pistils in all flowers. WESTERGAARD's "European" tetraploid XXXXY plants were mainly male, with only a small pistil in some flowers.

A hermaphroditic type of extra chromosomal nature has been described by many investigators. The most complete description is that given by BAKER (1947). Hermaphrodites of this type arise when an originally female plant is infected by the smut *Ustilago violacea*. On account of this, stamina of normal size develop, the anthers, however, are filled with violet spores. The pistils of these flowers are smaller than the pistils of female flowers.

In this connection the experiments of LÖVE & LÖVE (1945) should be mentioned. They stated to have induced staminodia in female flowers and pistils in male flowers with testosterone and oestrone respectively.

Hermaphrodites of unexplained nature were described by CORRENS (1924). He obtained these hermaphrodites by storing pollen, before pollination took place, for one week to four months over soda lime. Viability of the pollen decreased rapidly. The viability of the seeds dropped to one third after pollination with 47 days old pollen, and to zero after pollination with pollen stored for 120 days. From the sex ratios among the gerontogones, as he named the offspring, COR-RENS concluded that gametes with a Y-chromosome were more resistant to storage than the gametes with an X-chromosome. Selfing the hermaphrodites resulted unexpectedly in a female offspring, though sometimes also a single hermaphrodite appeared. Even part of the male gerontogones raised female offspring only.

The nature of the hermaphrodites found by HERTWIG & HERTWIG (1922) in *M. dioicum*, and of the somatic and genetic hermaphrodites obtained by SHULL (1910, 1911), is also not entirely clear. This may partly be due to the fact that at the time of these investigations the sex chromosomes were not yet discovered in *Melandrium*.

Many cytological investigations in *Melandrium* have been carried out, (for a review see LÖVE, 1944). BLACKBURN (1923) and WINGE (1923) independently of each other detected the sex chromosomes in *M. album*. They noticed that male plants had a heteromorphic pair of sex chromosomes and that in female plants the sex chromosomes were of equal size.

BLACKBURN (1924) regarded the largest chromosome in the male plant as the Y-chromosome. However, results of WINGE and others wrongly persuaded her that the largest one was the X-chromosome (BLACKBURN, 1929). WARMKE & BLAKESLEE (1939) obtained XXXY male plants by crossing 4n ♀ (XXXX) × 4n ♂ (XXYY). They found three short chromosomes and one long chromosome in the male plants. This proved that the single large chromosome must be the Y-chromosome.

WESTERGAARD (1940) obtained the same results with his tetraploid material. His results definitely indicated the identity of the X-chromosome and the Y, as he demonstrated that the Y-chromosome is metacentric and the X-chromosome submetacentric.

In 1946 WESTERGAARD described the results of his refined genetical and cytogenetical analysis of three very important aberrant Y-chromosomes (Fig. 1) selected by him among the offspring of triploid material.

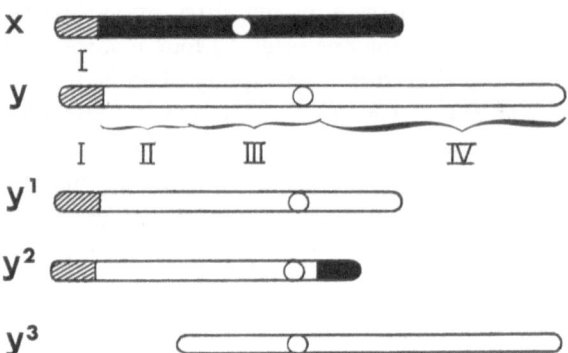

Fig. 1. Normal X- and Y-chromosomes and 3 incomplete Y-chromosomes obtained by WESTERGAARD (1946). I, homologous parts; II, ♂ promoting part (M₇); III, ♂ promoting part (M₁); IV, ♀ suppressing part. XY, ♂; XXY¹, ☿; XXY², ☿; XXY³, ♂ sterile. (After WESTERGAARD, 1946, 1948, 1958).

About half ot the non-homologous arm of the Y-chromosome is lacking in Y¹ and Y². A small distal part of the differential arm of the X-chromosome should be present in Y² instead.

In combination with some X-chromosomes, Y¹ and Y² gave rise to hermaphrodite plants. Obviously genes were absent that normally suppress the development of the pistil. In Y³ half the other arm of the normal Y-chromosome was absent, as a consequence of which only sterile male plants developed. The stamina degenerated after meiosis had been completed. The lacking part must therefore contain one or more genes necessary for the normal development of the stamina. Since for the development of normal male plants the whole Y-chromosome should be present WESTERGAARD deduced from his results that the part next to the centromere of the Y-chromosome must contain important genes necessary for the early steps in the development of the stamina.

From these and other data WESTERGAARD postulated the following mechanism of sex determination in *Melandrium*.

Dispersed over all chromosomes male (*M*) and female (*F*) promoting genes (the basic sex genes) should be present, which control the subsequent steps in the development of stamina and pistils respectively.

A trigger mechanism will switch on the male or female promoting genes. This trigger consists of female suppressor and male promotor genes, absolutely linked to each other in the differential part of the Y-chromosome. In the absence of the Y-chromosome only female flowers, and in the presence of this chromosome, only male flowers can be formed.

This mechanism has very important implications for the problem of evolution of dioecism. Since many, if not all, cases of dioecism in plants may be explained by the same principle, or by a principle evoluated from the type found in *Melandrium*.

Our experiments in *Melandrium* were undertaken because several questions remained open.

The role of the X-chromosome remains still somewhat obscure.

It must be realized that the only direct evidence for the existence of male promoting genes in the Y-chromosome is that the absence of half the partly homologous arm of the Y-chromosome, as found in one Y^3-chromosome, causes degeneration of the stamina after meiosis has completed.

The female suppressor genes have been placed in different arms of the Y-chromosome. WARMKE concluded that in his material these genes were situated in the partly homologous arm of the Y-chromosome. WESTERGAARD demonstrated definitely that in his "European" plant material the female suppressor element is in the differential arm.

The mechanism that releases the development of stamina in XX plants infected by *Ustilago* is not yet understood. In fact, this phenomenon seriously interferes with WESTERGAARD's hypothesis (GOLDSCHMIDT, 1955).

The nature of the hermaphroditic gerontogones of CORRENS is still obscure.

We will try to give an answer to at least some of these questions.

In the discussion a hypothesis will be developed on the mechanism of sex determination, which differs in several ways from the view given by WESTERGAARD (1940, 1948, 1958).

Material and Methods

Seed of *M. album* (Mill.) Garcke and *M. dioicum* (L.emend.) Coss. et Germ. was collected in a number of original habitats (NIGTEVECHT, 1966). The material was propagated in our experimental garden by cross breeding and inbreeding. Inbreeding was carried out by crossing female plants and male ones of the same family, a procedure which was repeated in a series of successive families.

The growing conditions and the crossing methods were the same as published before (NIGTEVECHT, 1966).

Flower buds of *M. album* and *M. dioicum* were fixated in Carnoy's fluid: 3 parts aethanol (100%): 1 part glacial acetic acid. After 24 hours we replaced the fixation fluid by 96% aethanol, a fluid which after another 8 hours was replaced by 70% aethanol. Subsequently the material was stored at $-20°C$ until squashes had to be made. Pollen mother cells were stained with 2% orceine in 60% acetic acid. About 20 minutes after squashing the preparations were placed upside down on a flat piece of dry ice ($-70°C$) for 2 minutes. Then the cover slip could be removed with a razor blade without damaging the cells. Via different mixtures of aethanol and glacial acetic acid (1 : 1; 3 : 1; 9 : 1) and a mixture of aethanol and xylol (1 : 1), the preparations were mounted with Canada balsam.

Before fixating root tips we kept the plants overnight at $+1°C$, so as to in-activate the spindle. The root tips were fixated in Carnoy's fluid and stained with basic-fuchsin (Feulgen) after hydrolysis in 1N HCl at 59°C for 8 minutes. During squashing we used plastic cover slips which subsequently were dissolved in aceton and replaced by glass cover slips mounted with Canada balsam as described by ÖSTERGREN & HENEEN (1962).

Results

GENETICAL DATA

A mutant Y-chromosome (Y^a) in M. album

After inbreeding for two generations, one hermaphrodite plant was noticed in a family (3G) descending from the *M. album* population R 255. The hermaphrodite had stamina and pistils in all flowers. The pistils of this plant had three to four stigmata instead of five (Fig. 2).

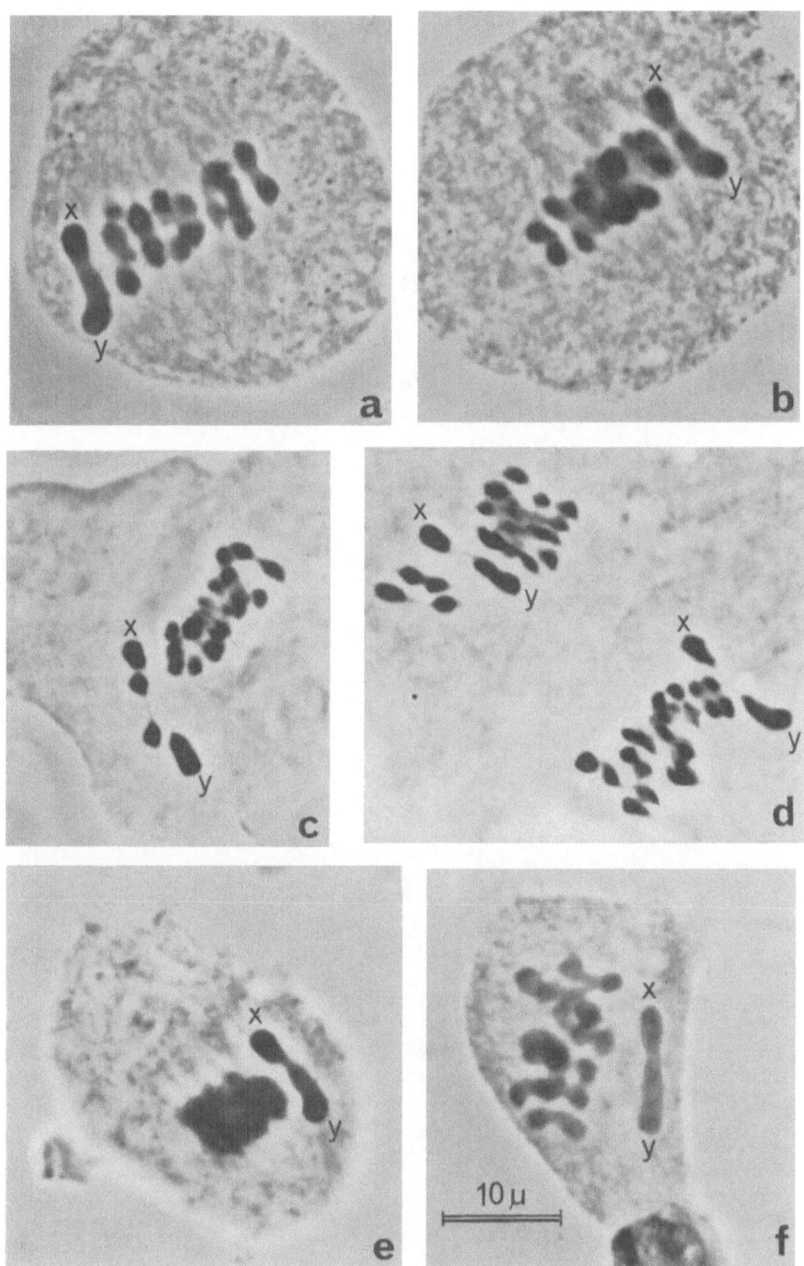

Plate 1. Meiosis, metaphase 1 of: a, *M. album* ♂; b, *M. album* ♀ XYᵃ; c, d, *M. album* ♀ gerontogones; c, 390 J; d, 387 J; e, *M. dioicum* ♂; f, *M. dioicum* ♀ (autosomal). (Orcein stain, phase contrast).

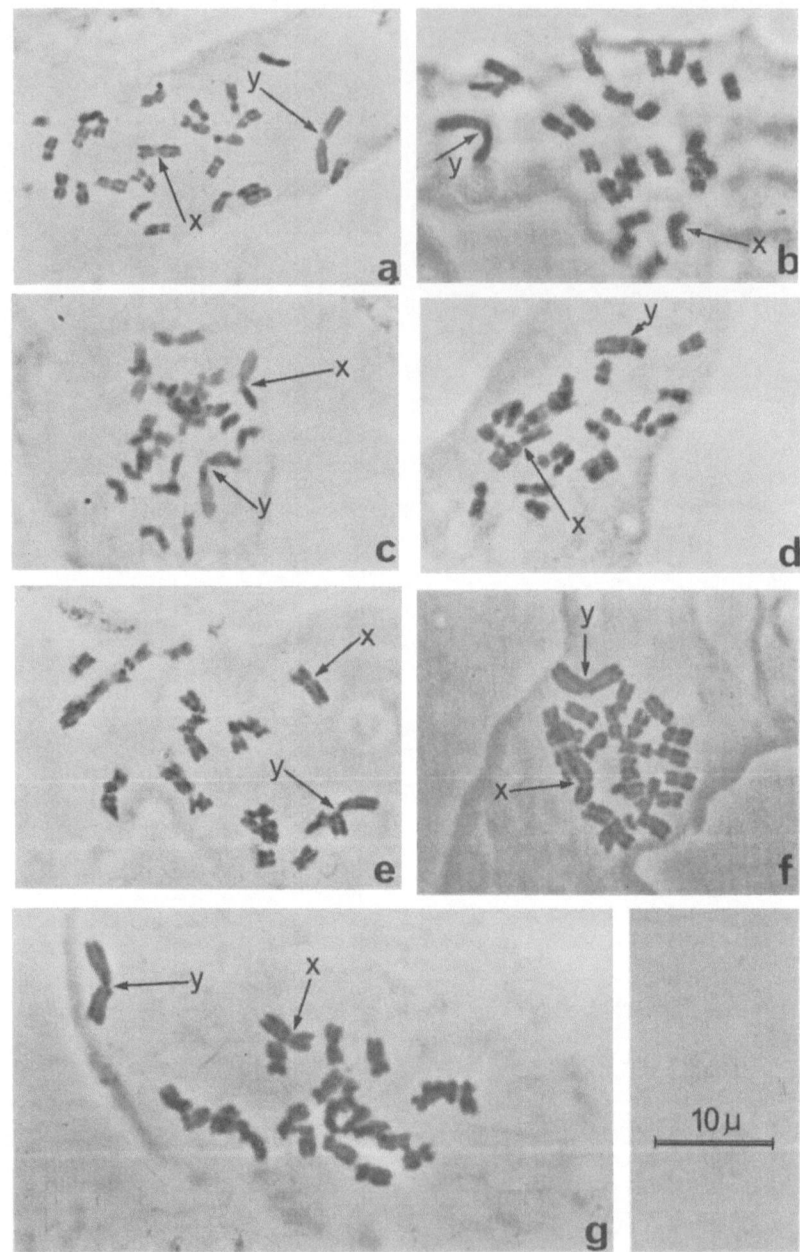

Plate 2. Mitosis: a, *M. album* ♂; b, *M. album* ♀ XY*ᵃ*; c, d, e, *M. album* ♀ gerontogones; c, 387 J; d, 388 J; e, 390 J; f, *M. dioicum* ♂; g, *M. dioicum* ♀ (autosomal). (Arrows indicate centromeres; Feulgen, phase contrast).

The fertility of the pistils was low, many pollinations with pollen from various families yielded no result. Some seeds could be obtained after selfing (196 H_1, Table 1) and after open pollination (196 H_2). The pollen of the hermaphrodite was quite fertile as seedsetting after pollination appeared to be normal (195 H). The stamina were less well developed than in normal male plants (Fig. 2).

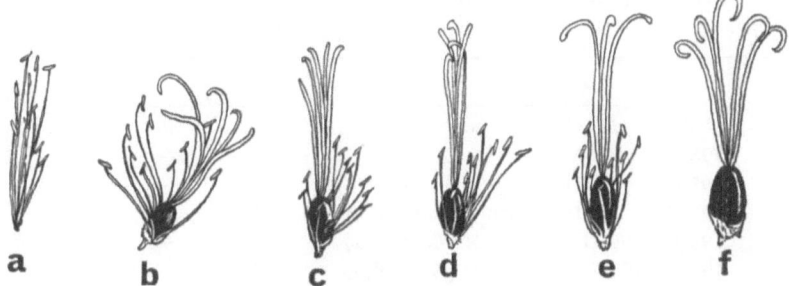

Fig. 2. Flowers of *M. album*, petals and calyx removed: a, ♂; b, c and d ♀ gerontogones; b, 387 J; c, 388 J; d, 390 J; e, 195 H, ♀ XY^a; f, ♀. (Natural size).

The appearence of one single hermaphrodite in family 3G could be the result of a new mutation, probably of a Y-chromosome, as the descendants with the Y-chromosome of 3G were hermaphrodites. The male individuals observed after open pollination in 196 H_2 are easily explained by cross pollination. On the other hand, the hermaphrodites noticed thus far, might also have been a product of inbreeding, because of which autosomal genes might have been accumulating that counter balance the female suppressing effect of a normal Y-chromosome. Outbreeding will of course break up such a special autosomal constitution, as a consequence of which, the normal situation with females and males will be reestablished.

Crosses (200 J, 199 J) were carried out with an unrelated diploid *M. album* plant and a tetraploid *M. album* individual (Table 1). Only females and hermaphrodites were obtained, also in the following generation (170 K, 172 K, 173 K). The variation in size of the pistils was the same as in the inbreeding material (195 H, 196 H, 195 J, 196 J, etc.), ranging from narrow pistils with two stigmata and larger pistils with four stigmata. Rarely were small pistils with one stigma and large pistils with 5 stigmata met with. In the triploid offspring about the same variation was noted, the pistils and stamina being a little larger than in the diploids.

We may conclude that the gene(s) in the Y-chromosome denominated by WESTERGAARD as female suppressors are absent, or have been mutated into a less effective state.

TABLE 1

CROSSES MADE WITH THE HERMAPHRODITE FOUND IN *M. album* (3 G) AND ITS HERMAPHRODITE OFFSPRING

		♀	♂	⚥
3 G	*M. album* ♀ × ♂	52	45	1
195 H	$3GIV_{28}$♀ × $3GII_{30}$⚥	47	—	39
196 H_1	$3GII_{30}$⚥ × $3GII_{30}$⚥	11	—	3
196 H_2	$3GII_{30}$⚥ open pollinated	39	20	24
195 J	$195HIII_{18}$♀ × $195HIV_{12}$⚥	37	—	58
196 J	$195HII_2$♀ × $195HIV_{23}$⚥	49	—	51
199 J	$199HIII_{15}$♀(4n) × $195HIV_{15}$⚥	71	—	45
200 J	$299HI_{16}$♀ × $195HIII_{25}$⚥	60	—	58
102 K	$196JIII_{22}$♀ × $196JIII_{21}$⚥	10	—	10
103 K	$196JIII_{21}$⚥ × $196JIII_{21}$⚥	28	—	21
172 K	$200JIII_{17}$♀ × $200JI_{21}$⚥	53	—	48
173 K	$200JI_{21}$⚥ × $200JI_{21}$⚥	63	—	52
170 K	$199JIII_{14}$♀(3n) × $199JIV_{17}$⚥(3n)	14	—	3

Both after inbreeding and after outbreeding with a non-related diploid female plant and a tetraploid one (200 J, 199 J), only female plants and hermaphrodites have been obtained. Male plants arose only after open pollination (196 H_2).

Mutation induction in the Y-chromosome of M. album and M. dioicum, as a result of pollen storage

For his experiments on the influence of age of the gametes on the sex ratio of *Melandrium*, CORRENS (1918–1924) stored pollen in desiccators over $CaCl_2$, H_2SO_4 concentrated and soda lime. With the latter method he had obtained seedsetting with pollen up to 110 days old. Pollen stored over $CaCl_2$ and H_2SO_4 kept its fertility no longer than 21 days. CORRENS assumed soda lime to be best because of less complete desiccation of the pollen and because no traces of SO_2 or HCl could be formed.

For our experiments we chose the following non-aggressive desiccants which cause different relative humidities of the air within a closed space: I, $Na_2CO_3 . 1H_2O + Na_2CO_3 . 10H_2O$, relative humidity

0.72; II, $Na_2CO_3 . 1H_2O$, rel.h. 0; III, Glycerol, rel.h. 0; IV, Glycerol 96% + H_2O 4%, rel. h. 0.15.

We stored pollen of two male plants for 1 to 8 weeks in desiccators. Already after one week of storage over desiccant I, the fertility of the pollen dropped to zero. With the other methods seedsetting was obtained. As no difference in effect between the other methods could be noticed, the data were pooled (Table 2). A decrease of viability of

TABLE 2

CROSSES MADE WITH POLLEN OF *M. album* STORED IN DESICCATORS AT 20°C FOR 0 TO 8 WEEKS

Storage time of pollen in weeks	Flowers pollinated	Capsules obtained	Viable seeds per capsule	♀	♂	♀̣	Total	Sex ratio ♀/♂
0	10	9	120	150	75	—	225 *)	2.0
1	15	7	24	118	51	1	170	2.3
3	48	14	15	144	58	—	202	2.5
5	29	8	18	121	24	—	145	5.0
8	57	24	3.4	61	21	—	82	2.9

Fertility of the pollen decreased rapidly. The increase of the sex ratio suggests that pollen with an X-chromosome might be slightly more resistant to the storage than pollen with a Y-chromosome. One hermaphrodite was obtained as a result of the storage of the pollen.

*) Only part of the seed sown.

the pollen with increasing time of the storage, was demonstrated. The increase of the percentage of male offspring as noted by CORRENS after pollination with old pollen, did not occur in our material. So the hypothesis of CORRENS that gametes with a Y-chromosome are more resistant to storage than gametes with an X-chromosome, could not be supported. On the contrary, as the sex ratio became even higher, while the certation effect must be weakened the gametes with an X-chromosome would be even more resistant than the gametes with a Y-chromosome.

We obtained one hermaphrodite with fairly well-developed pistils and stamina. With pollen of this plant seedsetting was obtained after pollinating two non-related diploid females (144 J, 262 J) and one tetraploid female (261 J). Also selfpollination produced seed (145 J).

These crosses with our hermaphrodite gave the same remarkable result as the crosses carried out by CORRENS with his hermaphrodites, since we obtained female offspring only (Table 4). CORRENS (1924) noticed that also part (¼th) of the crosses between female gerontogones and male gerontogones yielded exclusively female offspring. In our crosses this did not occur (Table 4).

In order to obtain more hermaphrodites of this type a second experiment was carried out. We collected pollen of two *M. album* plants (209 H (R 255), R 327) and of three plants of *M. dioicum* (R 330). The pollen was first kept in small open tubes in desiccators with the same four desiccants as used in the first experiment. After a fortnight, part of the tubes was sealed with paraffin wax. Of these some were placed at a temperature of + 3°C, others at −20°C. The unsealed tubes were kept at room temperature (ca 20°C) in the desiccators. Nine weeks after collecting the pollen, pollinations were carried out. All the pollen from one tube was used within three hours after opening.

As female partners, vigorous plants from various families were chosen. Each female plant of *M. album* was pollinated with pollen

TABLE 3

CROSSES CARRIED OUT WITH POLLEN OF *M. album* AND *M. dioicum* STORED FOR 9 WEEKS AT DIFFERENT TEMPERATURES

	Total	♀	♂	⚥
M. album				
20°C	212	105	104	3
3°C	506	287	214	5
− 20°C	164	67	97	—
			415	8
M. dioicum				
20°C	175	148	26	1
3°C	371	249	121	1
− 20°C	286	197	88	1
			235	3

In *M. album* 8 hermaphrodites and in *M. dioicum* 3 hermaphrodites with stamina and pistils in all flowers were obtained as a result of the storage of the pollen.

TABLE 4

GERONTOGONIC *Melandrium* PLANTS CROSSED TO GERONTOGONIC OR (IN SEVEN CROSSES†) TO NORMAL

				♀	♂	⚥
144 J	*M.a.* 295 H I 14	♀ × *M.a.*	25 H I 1 ⚥	69	—	—
145 J	,, 25 H I 1	⚥ × ,,	,,	38*)	—	—
261 J	,, 27 H II 18† (4n)	♀ × ,,	,,	110	—	—
262 J	,, 295 H II 12†	♀ × ,,	,,	117	—	—
51 K	,, 374 J I 3	♀ × ,,	374 J IV 8 ⚥	—	—	—
52 K	,, 374 J IV 8	⚥ × ,,	,,	—	—	—
87 K	,, 45 J III 1†	♀ × ,,	390 J VII 4 ⚥	76	—	—
182 K	,, 387 J I 11	⚥ × ,,	387 J I 11 ⚥	23	—	—
186 K	,, 45 J I 29†	♀ × ,,	,,	67	—	—
33 M	,, 19 L I 3†	♀ × ,,	,, much pollen	80	—	—
34 M	,, ,, †	♀ × ,,	,, few pollen	74	—	—
184 K	,, 377 J VIII 5	⚥ × ,,	377 J VIII 5 ⚥	3	—	—
187 K	,, 45 J I 29†	♀ × ,,	,,	67	—	—
53 K	*M.d.* 394 J V 1	⚥ × *M.d.*	394 J V 1 ⚥	124	—	—
54 K	,, 394 J IV 7	♀ × ,,	,,	123	—	—
274 J	*M.a.* 25 H I 10	♀ × *M.a.*	25 H I 9 ♂	8	13	—
275 J	,, 25 H II 4	♀ × ,,	25 H III 4 ♂	19	22	—
276 J	,, 25 H I 8	♀ × ,,	25 H I 3 ♂	3	11	—
277 J	,, 23 H I 9	♀ × ,,	23 H IV 2 ♂	14	18	—
282 J	,, 24 H I 1	♀ × ,,	24 H II 2 ♂	10	15	—
283 J	,, 24 H IV 1	♀ × ,,	24 H II 2 ♂	17	20	—
286 J	,, 20 H I 1	♀ × ,,	20 H I 5 ♂	26	12	6**)
18 K	,, 398 J VII 12	♀ × ,,	398 J VII 8 ♂	52	35	3**)
19 K	,, 398 J II 14	♀ × ,,	398 J III 4 ♂	55	41	2**)
56 K	,, 374 J I 12	♀ × ,,	374 J I 9 ♂	111	5	—
70 K	,, 275 J I 6	♀ × ,,	275 J I 2 ♂	45	51	—
71 K	,, 275 J IV 3	♀ × ,,	275 J VI 3 ♂	50	47	—
57 K	*M.d.* 394 J I 5	♀ × *M.d.*	394 J II 2 ♂	40	65	8**)

The hermaphrodites obtained in *M. album* and *M. dioicum* after pollination with old pollen yield female offspring only, both after selfing and in crosses when used as male parents. Crosses between male plants and female ones obtained after pollination with old pollen, yield female offspring and male, and a few hermaphrodites.

*) Including two plants with long staminodia.

**) These ⚥⚥ were androhermaphrodites.

collected from the two male plants. In *M. dioicum* each female was crossed with the three male individuals. Just as in the first experiment the pollen stored over $Na_2CO_3 . 1H_2O + Na_2CO_3 . 10 H_2O$, had lost its fertility completely. In table 3 our data are arranged according to the temperature during storage.

In *M. album* eight hermaphrodites, and in *M. dioicum*, three hermaphrodites were obtained with stamina and pistils in all flowers (Fig. 2). When these hermaphrodites were selfed or used as male partners for crosses, the offspring consisted of female plants only (Table 4, 51 K to 54 K). This can not be the consequence of an extreme certation effect, because half the female gametes should have an X-chromosome and the other half a Y-chromosome. Therefore, self pollination with male gametes with an X-chromosome and other male gametes with a mutated Y-chromosome should yield at least equal numbers of XY and XX individuals, even with complete elimination of the male Y gametes as a consequence of a strong certation. Moreover, after cross pollination with less pollen (Table 4, 34 M) also female plants only were obtained.

The fertility of the pollen of the hermaphrodites was low. Pollen preparations showed, on the average, less than 20 percent viable pollen.

Obviously the part of the Y-chromosome indicated by WESTER-GAARD as the female suppressor had mutated or was lost.

Our cytological data will demonstrate that the latter presumption is true.

As is shown in Table 3, in *M. album* 8 out of 423 Y-chromosomes (1.9%) mutated. In *M. dioicum* 3 out of 238 (1.2%) mutated.

An autosomal hermaphrodite in M. dioicum

Some crosses (Table 5, 25 E, 188 J, 222 J) carried out in different populations of *M. dioicum* (R 261, R 401, R 436), yielded one single androhermaphrodite each among normal females and males. Many flowers of these androhermaphrodites had stamina only, in others also a thread-like pistil was present. In a few flowers stamina and a narrow pistil with one or two stigmata were observed (Fig. 3). These pistils developed mainly in the first flowers of the inflorescenses. The capsules of these flowers contained at most 15 viable seeds. Selfpollination (Table 5, 46 F) of the androhermaphrodite of family 25 E, yielded

TABLE 5

CROSSES WITH HERMAPHRODITES IN *Melandrium dioicum*

			♀	♂	☿
222 J	*M. dioicum*	R 436 ♀ × R 436 ♂	55	22	1
188 J	,, ,,	R 401 ♀ × R 401 ♂	35	20	1
3 K	,, ,,	188 J ♀ × 188 J ♂	75	24	3
27 L	,, ,,	3 K ♀ × 3 K ☿	57	53	10
25 E	,, ,,	R 261 ♀ × R 261 ♂	104	88	1
36 F	,, ,,	25 E ♀ × 25 E ☿	35	14	23
46 F	,, ,,	25 E ☿ × 25 E ☿	23	—	19
139 G	,, ,,	46 F ♀ × 46 F ☿	5	9	3
73 G	*M.d.* × *M. (d × a)*	46 F ♀ × 4 F ♂	55	30	5
74 G	,, ,,	36 F ♀ × 30 F ♂	33	32	7
24 J	*M(d × (d × a))* × ditto	73 G ☿ × 73 G ☿	19	—	12
1 K		24 J ☿ × 24 J ☿	19	2	24
3 J		73 G ♀ × 74 G ☿	14	3	7
6 J		73 G ♀ × 73 G ☿	68	3	17
77 G	*M.a.* × *M.d.*	198 F$_1$ ♀ × 46 F ☿	81	14	1
78 G	,, ,,	198 F$_2$ ♀ × 36 F ☿	54	30	—
119 H	*M(a × d)* × ditto	77 G ♀ × 77 G ♂	106	24	—
122 H		78 G ♀ × 78 G ♂	76	42	—
123 H		78 G ♀ × 78 G ♂	84	37	7
155 H		78 G ♀ × 78 G ♂	67	30	1

In crosses (222 J, 188 J, 25 E) carried out in three different *M. dioicum* populations an androhermaphrodite arose. Inbreeding raised the frequency of the androhermaphrodite. No abnormal Y-chromosome could be involved, (see 73 G, 74 G, 77 G, 78 G). These androhermaphrodites appeared to be of autosomal nature.

23 female plants and 19 androhermaphrodites. In some of these androhermaphrodites a fair percentage of the flowers had stamina and a narrow pistil with two stigmata. In these plants after open pollination quite a number of capsules were noticed with ten to twenty seeds. On the other hand, in most of the androhermaphrodites of 46 F only with difficulty could a capsule with a few seeds be detected. A cross (36 F) between the androhermaphrodite of 25 E, and a female plant of the same family yielded a progeny consisting of 35 females, 14 males and 23 androhermaphrodites. Actually in this family the hermaphroditic character, on the whole, showed a lower expressivity than in family 46 F originating through selfing. In the course of se-

lection for larger pistils carried out during eight subsequent generations, (crosses not mentioned in the table) the expression of the herma-phroditic character became gradually more pronounced. Pro plant the percentage of hermaphroditic flowers increased until there were flowers of this type only. Simultaneously the pistils became larger and the number of stigmata increased to five pro pistil. In herma-phroditic flowers with pistils of a comparable size to those present in female flowers, the stamina appeared to be less well developed (Fig. 3).

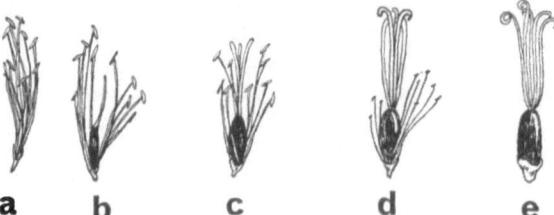

Fig. 3. Flowers of *M. dioicum*, petals and calyx removed: a, ♂; b, c and d ⚥ with 1, 3 and 5 stigmata respectively; e, ♀. The gynoecium in d equals the gynoecium of a normal ♀ flower (e), the stamina reduced in size (Natural size).

We carried out various crosses to investigate whether the herma-phroditic nature of these plants was determined by genes in the sex chromosomes or by genes present in the autosomes.

The Y-chromosome of the hermaphrodites could not be responsible because crosses (77 G, 78 G) with unrelated female plants gave rise to 114 females, 44 males and only one androhermaphrodite. As all 44 males, like the only androhermaphrodite, contained the Y-chromo-some from the hermaphroditic parent, this chromosome could not be responsible for the hermaphroditic nature of this XY plant.

So when the Y-chromosome had no weaker suppressing effect on pistil formation than normally, the X-chromosome could promote more strongly the formation of a pistil than normally. If this were so, then in the families 77 G and 78 G no androphermaphrodite should be observed, since the presumed abnormal X⚥-chromosome would be present in heterozygous condition in the female plants of these families (XX × X⚥Y → XX⚥ + XY). In the following generation half of the individuals with a Y-chromosome would have the abnormal X⚥-chromosome and as a consequence they should be hermaphrodite. We did not note anything of the sort (Table 5). In 77 G one androhermaphrodite appeared unexpectedly. In the families of the following generation (119 H, 122 H, 123 H, 155 H) we came across 8 andro-hermaphrodites, 133 males and 333 females. In these families equal numbers of hermaphrodites and of males were expected if the X-chromosome was responsible. Therefore no abnormal X-chromosome could be involved.

As neither genes in the Y-chromosome nor genes in the X-chromosome appeared to be responsible for the development of the pistils in the XY individuals, autosomal genes had to be. This was already suggested by the gradual shift towards femaleness during our selection. As this pointed to the fact that a fair number of genes must be involved, with each a small effect on the development of the pistil.

It is remarkable that we noticed hermaphrodites of this type in *M. dioicum* only, where they originated in the first inbreeding generation of three independent populations from different origin. This is the more striking since we analyzed of *M. album* about 60.000 XY individuals and only about 10.000 XY plants of *M. dioicum*.

CYTOLOGICAL DATA

The three types of hermaphrodites I found in my material showed a normal conjugation of the X- and Y-chromosomes (Plate 1). The hermaphrodites that originated from pollinations with old pollen, displayed an abnormal meiosis in part of the pollen mother cells since bridges were observed in anaphase I and anaphase II.

In 30 mitotic cells pro plant I measured the length of the arms of the Y-chromosome and the total length of the X- and Y-chromosomes in metaphase plates. Normal Y-chromosomes are metacentric (Plate 2) both arms having the same size. If one arm is missing a part, the relative size of this part will show up by counting (l.a. — s.a.)/2 l.a. since the result represents the part of the Y-chromosome that has been missing. A comparison of our data (Table 6) demonstrates no difference between the Y-chromosomes of an XY^a hermaphrodite (200 J) an autosomal *M. dioicum* hermaphrodite (3 J) and a normal male plant (45 J). The hermaphrodites obtained after pollination with old pollen have Y-chromosomes where a distinct part is missing.

We assume that the loss of these parts of the Y-chromosomes has been caused by a starvation process during the storage of the pollen.

TABLE 6

MEASUREMENTS OF SEX-CHROMOSOMES

Plants investigated	Mean $\dfrac{\text{l.a.}(Y) - \text{s.a.}(Y)}{2\,\text{l.a.}(Y)}$	Mean $\dfrac{l(Y)}{l(X)}$
45 J ♂ *M. album*	0.01 ± 0.0004	1.42 ± 0.04
3 J ☿ *M. dioicum*	0.01 ± 0.0003	1.46 ± 0.05
200 J ☿ *M. album* XY[a]	0.01 ± 0.0003	1.49 ± 0.04
377 J ☿ *M. album* gerontogone	0.14 ± 0.003	1.25 ± 0.03
387 J ☿ *M. album* gerontogone	0.15 ± 0.004	1.21 ± 0.04
388 J ☿ *M. album* gerontogone	0.25 ± 0.008	1.07 ± 0.03
390 J ☿ *M. album* gerontogone	0.20 ± 0.009	1.12 ± 0.03

Mean of 30 cells per plant demonstrating a distinct difference between the long arm (l.a.) and the short arm (s.a.) of the Y-chromosomes of gerontogonic hermaphrodites. The two other hermaphrodites, like the normal male plant, have metacentric Y-chromosomes.

The actual fragmentation happened most probably at pollen mitosis during germination of the pollen. The irregular meiosis in some of the pollen mother cells of the resulting hermaphrodites is partly due to the fragmentation of some autosomes that may have taken part in translocations. Though the very low percentage of viable pollen (about 10%) cannot be explained by this cause alone. This low fertility might mainly be attributed to a poorer development of the stamina.

Discussion

As a consequence of his investigations in *Melandrium*, WESTER-GAARD (1940, 1948, 1958) has proposed a hypothesis about the evolution of dioecism in plants.

Since his pioneering work on this subject several new facts have become available which make it desirable to revise the hypotheses about the mechanism of sex determination and the evolution of dioecism.

WESTERGAARD (1958) regretted the fact that so few physiological data on sex determination were within reach, because he assumed these a necessity for a final solution of the problem.

We are of the opinion that the distribution of the female flowers and male ones in monoecious plants is of importance. The experiments of recent years on the influence of auxins and related substances on the development of androecium and gynoecium are likewise of interest. Particularly since more data are becoming available about the influence of hormones on gene action.

These new facts and the results of our own investigations satisfied us that the sex-determining genes, i.e. the genes that lead to the development of a gynoecium and/or an androecium, create a physiological situation such that a change of this situation always acts in favour of the development of one of the sex organs and consequently against the other. This situation resembles essentially the F/M balance of GOLDSCHMIDT (1955).

An increasing number of data have become available which suggests that this physiological situation is mainly characterized by the net auxin activity. An increase being in favour of the development of the gynoecium and of disadvantage to the development of the androecium. A decrease having the opposite effect. This naturally genetically determined physiological mechanism has its implications for the explanation of the sex determining mechanism in *Melandrium* and for the theory about the evolution of dioecism. Furthermore, we noticed a physiological relation between the origin of dioecism and apomixis.

DIOECISM AND MONOECISM

Dioecism is comparatively rare in plants. YAMPOLSKI & YAMPOLSKI (1922) noticed that in phanerogams 70% of the genera are wholly hermaphrodite and 5% wholly dioecious. LEWIS (1942) found that of the species of the British flora 92% was hermaphrodite, 5.4% monoecious and 2% dioecious.

Further data of YAMPOLSKI & YAMPOLSKI (1922) and LEWIS (1942) demonstrate that dioecious species are much more frequently associated with monoecious ones than with hermaphrodites, suggesting that monoecious species and dioecious ones have arisen from hermaphrodite species through related processes.

In monoecious types, flowers of different sex arise on the same plant. In dioecious species female flowers and male flowers are formed on different individuals with different genotypes. The formation of a female flower and of a male flower depends, like the formation of leaves, shoots etc., on processes of determination and differentiation which in monoecious species are controlled by but one genotype. These processes are affectable by the outer milieu and by changes in the inner milieu caused by the continuous development of the plant.

In *Cucurbita pepo* a shift from maleness towards femaleness proceeds during the development of the plant. In succession are formed small sterile male flowers, fertile male flowers, alternatively female flowers and male ones, large female flowers alternated sometimes with small sterile male flowers and, finally, a parthenocarpic female flower (NITSCH et al., 1952).

The shift towards femaleness in *Cucurbita pepo* is caused by the changing outer milieu in the course of the season and probably also by changes of the inner milieu as a result of the continuous development of the plant.

This last factor is solely responsible for the transition from maleness to femaleness, as normally occurs in *Arisaema triphyllum*. When this Araceae starts flowering, the first year only male flowers are present. In following years only female flowers arise. Obviously changes of the inner milieu of the plant control the changing expression of the sex determining genes.

Experiments (SCHAFFNER, 1922, 1925) have demonstrated that accumulation of assimilates especially in the continuous growing corm, the axis on which the flowers develop, is responsible for the shift to femaleness.

It is easily understood that in a monoecious plant male flowers need for their creation another inner milieu than female flowers. By what are these respective milieus characterized? GOEBEL noticed in 1908 already that *Funaria*, fern prothallia, *Begonia* and other plants when grown under poor nutritional circumstances did not reach the female phase, but only the male one. It has been demonstrated in monoecious species, that, on the whole, circumstances that stimulate vegetative growth promote the formation of female flowers and simultaneously inhibit the formation of male ones (HESLOP-HARRISON, 1957).

Besides these influences of environmental factors as nutrition, light, temperature, humidity and others it has been shown that a developing gynoecium in one flower affects the sex determining processes in other flowers. Removal of female flowers in *Cucumis sativus* leads to the formation of more female flowers instead of the expected male ones

(TIEDJENS, 1928). It is assumed that female flowers inhibit the forming of new female flowers but not of male ones. The latter exert no influence on other developing flowers, they do not suppress the development of a gynoecium in later flowers. Thus alternatively female flowers and male ones might be formed until, as for instance, in *Cucurbita pepo*, other influences cause a complete transition to femaleness.

Interesting in this respect is *Cleome spinosa*. In the inflorescence of this species zones with female flowers and zones with male ones alternate. Between two zones of pure female flowers and pure male ones there is a series of transitional forms in which, starting at the female side of the series, the gynoecium becomes gradually smaller and the androecium simultaneously larger. The series of flowers developing after the zone of male flowers displays a similar gradual transition to the following zone of female flowers. *Cleome* demonstrates clearly that an improvement of the inner milieu for the development of a gynoecium therefore makes, to the same extent, the inner milieu less suitable for the development of an androecium. In *Cleome* too the removal of female flowers leads mainly to the formation of female flowers.

The question arises which physiological principle under the influence of factors of the outer milieu and of a developing gynoecium might change the expression of the sex-determining genes.

The distribution of the flowers on *Procris laevigata* might give the clue to this problem (GOEBEL, 1913). This woody Urticaceae has its leaves in four rows on the stem. As a consequence of its habitat one side of the plant is turned to the light. The leaves on the side of the plant that is turned to the light are distinctly smaller than the leaves on the shady side. In the axils of the small leaves male inflorescences are developing, in the axils of the large leaves the female inflorescences are found.

In this instance neither a change of the milieu in the course of the season, nor an influence of developing gynoecia on the formation of gynoecia in later flowers, seems to be responsible for the distribution of male flowers and female but most probably the influence of one milieu factor is. The light seems to be the most important factor. It is known that light affects the amount of active auxin in the plant in such a way that in the stem more active auxin is present on

the shady side than on the other side of the stem. Therefore the distribution of female inflorescences and male ones in *Procris* might be
regarded as an illustration of the importance of auxins in the processes
of sex determination.

SEX DETERMINATION AND AUXIN ACTIVITY

The results of a number of experiments have demonstrated the
important role of growth substances in sex determination processes
(HESLOP-HARRISON, 1957). In the monoecious cucumber, *Cucumis
sativus* (LAIBACH & KRIBBEN, 1950) and in the pumpkin, *Cucurbita pepo*
(LAIBACH & KRIBBEN, 1951a, b) indol acetic acid (IAA) and also
naphtoxy acetic acid (NAA) do promote the formation of female
flowers at the cost of the development of male ones. Even in the
dioecious hemp, *Cannabis sativa*, HESLOP-HARRISON (1956) was able
to induce with NAA the development of female flowers instead of
male, in male plants.

HESLOP-HARRISON (1957) put forward the hypothesis that for
stamen growth and growth of the pistils different optimal auxin
concentrations are needed, the concentration promoting the maximal
growth of stamina being lower than the one that promotes pistil
growth best.

According to this hypothesis in certain flower primordia of *Cucumis*
the auxin concentration would be too high for the development of the
androecium, but suitable for the formation of the pistil. This developing pistil would lower the auxin concentration in some following flower
primordia so that there no pistil could be formed but an androecium
could. In flower primordia developing after the male flowers, gynoecia
can again be formed.

A flower can be taken as a shoot. For a developing shoot the so-
called "apical bud dominance" is a very important principle. This
principle implies that the apical bud inhibits the development of some
lower buds on the shoot. Removal of the inhibiting bud, releases the
development of lower buds. If, however, after the removal auxin is
supplied to the tip of the shoot the lower buds will remain dormant.
In the course of the development of the shoot the lower buds will
one day develop. The time of this event will depend on the strength
of the apical dominance and the distance between inhibiting bud and
the inhibited ones.

In a flower primordium the "apical dominance" could be exerted by the developing pistil. In a normal hermaphroditic flower this apical dominance would not inhibit the formation of stamina. However, a stronger apical dominance of a pistil might inhibit the development of the stamina in the same flower, thus giving rise to an unisexual flower.

As has been demonstrated the gynoecium might influence other developing flower primordia as well. Unlike in normal "apical bud dominance" the influenced flower primordia might be present above the inhibiting flower. This is no objection since LIBBERT (1954, 1955) has demonstrated an inhibiting effect of a bud of a side branch on the growth of the main stem in *Pisum sativum*. Administration of auxin to the growing point of the main stem neutralizes the inhibition. This inhibition by a bud on a side branch on the development of the main stem is comparable to the inhibition by a female flower on the development of newly formed flowers on a continuously growing stem. This inhibition can also be neutralized with the aid of auxin.

We are of opinion that the physiological mechanism that is involved in sex determination might be very similar to the physiological processes underlying the phenomenon "apical bud dominance". In what way the apical bud can have influence on the development of lower buds is not known. In one hypothesis it is assumed that an inhibitor is formed out of auxin or under the control of auxin (SNOW, 1940; LIBBERT, 1954, 1955). The concentration of this inhibiting substance would increase to a certain extent with the distance to the dominating apical bud.

In a second interesting hypothesis, the auxin itself is regarded as the inhibiting substance for the lower buds. High concentrations of auxin would inhibit the formation of vascular connections to the buds (THIMANN & SKOOG, 1934; THIMANN, 1954) thus impeding the supply of nutrients (GREGORY & VEALE, 1957).

CRAFTS (1961) put forward the hypothesis that active growing tissues cause a flow of nutrients towards itself which might cause a drainage of auxin from less active tissues. Whatever might be the right physiological explanation, the optimal auxin concentration for the development of a gynoecium appears to be higher than for the development of an androecium. Convincing in this respect are the experiments of GALUN et al. (1962). Cucumber flower buds that should have become male flowers, developed into female flowers when grown in vitro with IAA in the substrate.

In vivo auxin activity will not solely be determined by the presence of a special amount of auxin, but also by the possible presence of substances as anti-auxins and the like. A combination of these substances will result in a net auxin activity. This net auxin activity might be caused by a fraction of the total amount of auxin present at a certain point, nevertheless, this net activity is of importance only for the sex determination. Of interest in this respect are the investigations about the influence of gibberellins on the development of gynoecium and androecium. (SHIFRISS, 1961, 1964; PETERSON & ANHDER, 1960; MITCHELL & WITTWER, 1962).

In a monoecious plant like *Cleome spinosa* in certain flower primordia a high net auxin activity will be present as a consequence of which the pistil develops and the stamina are inhibited, perhaps by inhibition of the formation of vascular connections towards the stamina. In some following flower primordia the net auxin activity might be too low for a normal development of a gynoecium, perhaps on account of the formation of an inhibiting substance or because of a possible drainage of auxin by the developing gynoecium.

In pure hermaphrodite plants all flower primordia will allow the development of both gynoecium and androecium, the primordium being more or less balanced in this respect. For the origin of monoecism, this balance should be disturbed.

It is tempting to explain the origin of monoecism by an increase in the net auxin activity in a flower primordium by which the formation of the androecium in that flower might be inhibited, and the developing gynoecium possibly inhibits the development of gynoecia in following flowers. In dioecious species like *Melandrium*, in male individuals the conditions will be permanently similar to those in the male zone of *Cucurbita pepo* and in female plants as in the female zone of *Cucurbita*.

In an XX plant of *Melandrium* we thus might expect a higher net auxin activity than in XY plants, at least in the developing flower buds.

THE ORIGIN OF DIOECISM

In view of the variable expression of the sex-determining genes in monoecious species, it seems possible that there is a type of dioecism for which only the milieu factors are responsible. In animals an example is known. Fertilized eggs of the marine worm *Bonellia* develop to females when they become attached to the substrate. However, when they attach to a female individual, exclusively males come into existence. This example strongly resembles the inhibiting influence of female flowers on other developing flower primordia as demonstrated in various monoecious plant species. The Araceae, *Arisaema* displays a kind of pseudo dioecism as plants of different age might produce flowers of different sex.

For the existence of a stable dioecious plant species constant differences in the inner milieu of female plants and male ones seem to be required. A situation which can be accomplished by genic differences only.

Taking the auxin hypothesis for granted the net amount of auxin will be higher in the flower primordia of female plants than of male ones, these differences being genetically determined. The genes which increase the net amount of auxin and consequently promote the development of gynoecia will be denominated as + genes. Genes with an decreasing effect on the net auxin activity and as a consequence promoting the development of androecia, will be indicated as − genes. Naturally these + and − genes represent a very heterogeneous group of genes, as they might influence auxin production, auxin transport and also the production and transport of various auxin antagonists.

In hermaphrodite plants these + and − genes are present in such a relation that both an androecium and a gynoecium will develop in every flower. This more or less balanced condition might be disturbed by a change in the proportion of + and − genes. A slight increase in + genes might lead to the origin of gynohermaphrodites, slightly more − genes might give rise to the origin of androhermaphrodite plants. A further shift in either directions might cause the transition of these monoecious types into unisexual forms.

Examples of such transitions have been observed in *Zea mays* (EMERSON, 1932; JONES, 1932, 1934). The mutation *Sk* → *sk* (*silkless*) causes underdeveloped pistils in the female inflorescence, consequently the plants are male. In our terminology by this mutation a + gene has been changed into a — gene. Another mutation $Ts^2 \to ts^2$ (*tassel seed*) represents a change from — to + since pistils instead of stamina develop in the flowers of the male inflorescence.

The + effect of the ts^2ts^2 might be regarded as stronger than the − effect of the *sk sk* mutant since the former causes a sex reversion and the latter only a decrease in the expression of one sex. This appears also from the fact that when both mutations are combined in one individual, this is a ♀ plant. Quite a number of these + and — genes are known in *Zea mays*. It is understandable that a simple dioecious type as has been synthesized by Emerson and Jones in *Zea mays* *sk/sk* ts^2/ts^2 (♀), *sk/sk* Ts^2/Ts^2 (♂), will in nature, not last, after a possible spontaneous origin. Other mutations and also recombinations between + and — genes, present in heterozygous condition, will easily break up such a dioecious situation which is based on a single gene difference only.

Yet the condition as has been demonstrated by EMERSON and JONES might give the clue for the solution of the problem how a stable dioecious situation arises. It seems necessary that female genotypes and male ones should differ in more than one gene. Such a constant difference in more than one gene seems to be realizable only with the aid of absolute genetical linkage.

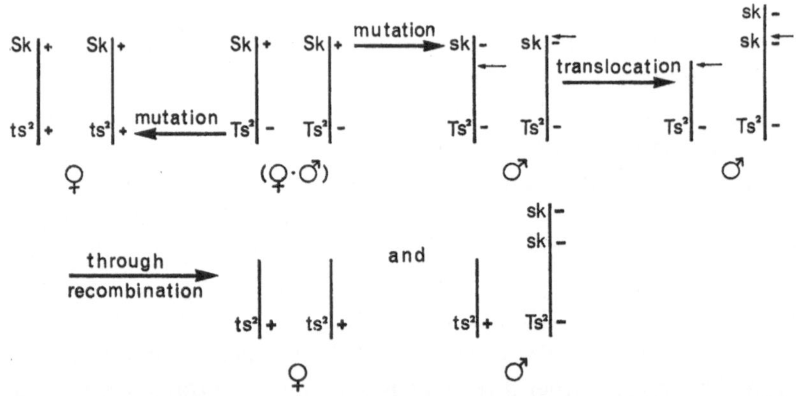

Fig. 4. Mutations of *Zea mays* are assumed to be linked so as to show a possible evolution of sex chromosomes, through an unequal distribution of + genes (relatively high auxin activity) and — genes (relatively low auxin acitivity) over a pair of originally homologous chromosomes. (Arrows indicate the break points involved in translocation).

If the two loci of Sk and Ts^2 were located in homologous chromosomes the two mutations could give rise to a $\frac{+\,+}{+\,+}$ female plant and a $\frac{-\,-}{-\,-}$ male one (Fig. 4). These genotypes will not make up a dioecious species. A population might arise consisting of male plants, females and monoecious individuals. A translocation in the $\frac{-\,-}{-\,-}$ male might give rise to the formation of a pair of chromosomes which are partly non-homologous (Fig. 4). In this combination two — genes are absolutely linked. Trough recombination various genotypes will be formed in a population. Among these genotypes $\frac{+}{+}$ females and $\frac{+}{-\,-\,-}$ males (Fig. 4) might be present. These genotypes selected from the population constitute a dioecious combination, provided the third — gene is absolutely linked to the other two — genes in the longest chromosome, for instance by an inversion.

Similarly, sex chromosomes might have originated by a transformation of two homologous chromosomes into partly non-homologous ones in such a way that in one chromosome especially + genes and in the other – genes have accumulated.

SEX DETERMINATION AND GENE ACTION

The development of sex organs will be controlled by genes. For the formation of a gynoecium and an androecium the same genes will to a certain extent be involved, like for instance genes with primary cell functions as cell division and the like. The different outcome of the activity of these genes will be due to differences in the duration of the action of the respective genes. Especially small differences at the start of the development might cause a considerable change in the further development.

We have argued that auxin plays an important part in sex determination, we might therefore ask what is the function of auxin in the correlation and integration of cell functions, in other words how can auxin, itself dependent on gene activity, influence gene activity so considerably.

The investigations of BEERMAN (1964) in the mosquito *Chironomus* might give a clue to this problem. A particular, low concentration of the pupation hormone ecdyson induces a gene to code messenger RNA, as a consequence of which, shortly after, other genes start coding RNA. At a higher concentration of ecdyson another gene is activated, also followed up by various other genes that depend for their action on the activity of this gene. Auxin instead of ecdyson could have such a regulating function.

It is known that gibberellic acid, like auxin called a phytohormone, directly acts on the formation of specific mRNA (VARNER & CHANDRA, 1964, 1965). In barley endosperm gibberellic acid promotes the formation of α- amylase via an increased production of specific mRNA. The outcome of experiments on *Pisum sativum*, *Avena sativa* and *Helianthus tuberosus* were to NOODÉN & THIMANN (1963) strong indications that "the locus of action of auxin in cell enlargement is on a nucleic acid system controlling synthesis of an essential protein". HAMILTON et al. (1965) demonstrated a stimulation of RNA synthesis in *Avena* coleoptiles by IAA.

Thus it is quite conceivable that auxin like gibberellic acid and like many animal hormones (DAVIDSON, 1965) might activate certain genes through the start of mRNA synthesis. Hormones could very well operate on the regulation of protein synthesis through a mechanism as described by JACOB & MONOD (1961). JACOB and MONOD are of the opinion that cell metabolites will influence the relation between a repressing form and a non-repressing form of the product of a so-called regulator gene (repressor). The repressor prevents the action of so-called structural genes, which are ultimately responsible for the structure of the proteins, by blocking the synthesis of mRNA on the DNA (JACOB & MONOD, 1961), or by blocking protein synthesis on the ribosomes in contact with mRNA (GRUBER & CAMPAGNE, 1965).

According to this concept, genes are (temporary) inactive because of the presence of a repressor. They become activated when the products of the respective regulator genes change to a non-repressing state. This change might be caused by a metabolite (inductor) synthesized in the cells with the repressed genes, or by a metabolite transported from elsewhere, as has been observed often in case of animal hormones.

Auxin could exert such a concentration dependent regulating influence on the action of sex-determining genes.

SEX DETERMINATION IN *Melandrium*

WESTERGAARD (1948, 1958) had distinguished two groups of sex-determining genes. The "sex-promoting or basic sex genes" which might be present in all chromosomes including the sex chromosomes, and which should be present also in hermaphrodite species and monoecious ones. Another group of genes forming a "trigger" mechanism would consist of so-called "sex deciding genes", which are present only in the differential segments of the sex chromosomes. This "trigger" would switch the "sex promoting genes" so that alternatively female sex organs or male ones would develop.

WESTERGAARD (1946) has regarded genes in the non-homologous arm of the Y-chromosome as representing the female suppressor element of the trigger. When a part of this arm of the Y-chromosome is missing pistils could develop next to the stamina in the flowers of an XY' plant.

When a part of the other arm of the Y-chromosome is missing the development of the stamina was arrested after meiosis had been completed. According to WESTERGAARD in this case a sex-promoting gene (M_7) was lacking, necessary at the end of the development of the stamina. Since in the absence of a complete Y-chromosome in XX plants no stamina will develop, WESTERGAARD has deduced, near the centromere of the Y-chromosome, the presence of at least one male sex-promoting gene indispensable for the early development of stamina.

We would suggest that it will be difficult to maintain a fundamental division in sex-promoting genes and sex-deciding ones. The formation of an androecium or a gynoecium is not comparable to the biosynthesis of an amino acid as it is done (WESTERGAARD, 1958). In morphogenetic processes leading to the development of different organs, as for instance an androecium and a gynoecium, partly the same genes will be involved, as for example genes for cell division. Other genes will give rise to differences in correlation and integration of the activities of the joint genes. These genes together will be responsible for the development of a gynoecium or an androecium, therefore they might be regarded as "sex-determining genes". We have already argued about the importance of the auxin activity for the development of the respective sex-organs. Consequently the sex-determining genes them-

selves must be involved in the regulation of the net auxin activity in
the developing flower primordia; an increase in the auxin activity
acting simultaneously female promoting and male suppressing, thus
acting as sex-promoting genes and sex-deciding ones at the same time.
A reduction of the net auxin activity will cause the opposite effect.

In case one would prefer to distinguish a "trigger" it should be
realized that this "trigger" consists of a number of the "basic sex
genes" unevenly distributed over two partly non-homologous chromosomes. Thus two different chromosome combinations can be formed
which give rise to the existence of two different auxin activities in the
respective genotypes.

Our conception is based on a number of facts. The genes on the
Y-chromosome that suppress the development of the pistil simultaneously promote the development of the stamina. In our XY'
material the fertility of the pollen has decreased markedly, in some
cases pollen is even fully sterile. Indeed, pictures of some of WESTER-
GAARD's hermaphrodites (1946) demonstrate the effect of such a
balance. Since a $42 A + XXXY^2$ plant shows well-developed stamina
and a rather small pistil, whereas another plant with the same Y^2-
chromosome $44 A + XXXY^2$ shows a well-developed pistil but small
stamina. Obviously the phenotypical differences have been caused by
the autosomes (A).

Also in normal XY plants a gynoecium might develop. This we
noticed in $M.$ $dioicum$ plants, where subsequently, with selection we
could obtain a gradual increase in the size of the pistils. Obviously
genes in the autosomes and in the sex-chromosomes of an XY individual normally bring about a condition in the flower primordia that
suppresses the development of a gynoecium and promote the development of the androecium. In $M.$ $dioicum$ it appeared possible to change
this condition by autosomal selection so that at last large pistils and
underdeveloped stamina are present in diploid XY plants.

The hermaphrodites found by HERTWIG & HERTWIG (1922) in $M.$
$dioicum$ are probably of the same type as our autosomal hermaphrodites. This may be concluded from their description of the phenotypes
and their extensive investigations. Also the so called somatic and
genetic hermaphrodites described by SHULL (1910, 1911) could be of
this type.

WESTERGAARD's idea that in the Y-chromosome so-called male-promoting genes would be present, necessary for definite steps in the development of an androecium, does not agree with the observations on female *Melandrium* plants infected by the smut *Ustilago violacea*. In the flowers of these XX plants stamina of normal size develop, the thecae being filled with smut spores. The pistils in these flowers are most of time strongly reduced. It seems particularly unlikely that this fungus could replace the effect of the various M genes if we assume these genes to take different specific steps in the development of the androecia. It would imply that the fungus could produce the specific products of M_1 and M_7 and at the same time imitate to a certain degree the activity of the so-called female suppressor genes. It is more reasonable to assume that *Ustilago* does not produce the specific products of all these genes, but that the smut influences the total effect of perhaps all the sex-determining genes, probably by lowering the net amount of auxin in the flower primordia.

Also in XX plants not infected by *Ustilago*, stamina might develop in some degree. Especially in families with general growth inhibitions we observed in XX plants staminodia sometimes up to half the length of normal stamina whereas the pistils were underdeveloped. Among the XX descendants obtained after selfing a hermaphrodite geron-togone some plants were noticed with stamina of nearly half the normal size. Since the X-chromosome of the hermaphrodite originated from a normal parent this chromosome could hardly be the causing factor. Moreover, all XX individuals in the progeny of the hermaphrodite were homozygous for the same X-chromosome. If this X-chromosome caused the development of the stamina all XX plants should have had stamina in their flowers. So autosomal mutations probably originated during the pollen storage were responsible for the development of the stamina.

DIFFERENT PROPORTIONS + GENES — GENES IN *M. album* AND
M. dioicum

Only out of *M. dioicum* could we select our autosomal herma-phrodites. It made no difference whether the Y-chromosome of *M. dioicum* or of *M. album*, introduced by repeated backcrossing, was present.

It may well be possible, however, that the X-chromosome of *M.*

dioicum has more or more critical + genes than the X-chromosome of
M. album, though it seems more probable that the difference is to be
found in the autosomes.

This difference between *M. album* and *M. dioicum* might have
caused the phenotypical differences between the tetraploid XXXXY
individuals of WARMKE (1946) and of WESTERGAARD. The former
being true hermaphrodite, the latter producing mainly pure male
flowers. It is quite possible that the tetraploids of WARMKE (1946)
were not of pure *M. album* origin. It is likely that also genes of *M.
dioicum* were present. BAKER (1948) stated that *Melandrium* was not
introduced into the United States of America as two pure species.
From the beginning already *M. album* would contain genes from *M.
dioicum*. Our investigations point to the fact that probably because of
autosomal genes the female suppressing effect of the Y-chromosome
can be overcome in *M. dioicum* more easily than in *M. album*. Consequently, in WARMKE's material 4 X-chromosomes in the presence of
at least some *M. dioicum* genes might overrule more easily the female
suppressing effect of a Y-chromosome, than the 4 X-chromosomes in
the certainly pure *M. album* material of WESTERGAARD might.

DIOECISM AND APOMIXIS

The sequence of the various flower types along the stem of *Cucurbita
pepo* is interesting in so far a transition from a male phase to a female
one culminates in the development of a parthenocarpic female flower.
According to our concept for the development of this flower a still
higher auxin activity would be required than for the preceding female
flowers.

It is in fact known that administration of growth substances in
various plant species like tomato and strawberry will cause the development of parthenocarpic fruits, (fruits developing without fertilization
NITSCH, 1965). OVERBEEK, CONKLIN & BLAKESLEE (1941) injected
growth substances into the ovaries of *Melandrium* and *Datura*. Fruits
and seed coats had been formed without fertilization. In *Datura* cells
of the inner integument formed a kind of pseudo embryos, these were
regarded as possible adventitious embryos. YOUNG (1943) has treated
a parthenocarpic cucumber variety with auxin. He once observed an

unfertilized embryosac with nuclear endosperm and a very small embryo.

These and other results (OVERBEEK et al., 1941) suggest that apomixis might require, like parthenocarpy, a high auxin activity, perhaps even higher than the parthenocarpic condition.

If we assume that apomixis is like parthenocarpy characterized by a higher requirement of auxin in the developing flower primordia than female flowers it is no longer surprising that apomixis frequently occurs in genera with related monoecious species and dioecious ones, and that many flowers of apomictic plants produce no or only sterile pollen. This is noticed in others in microspecies of *Taraxacum* where many apomicts might be regarded as physiological female plants (GUSTAFSSON, 1947).

We think that dioecism can develop via a variable monoecious population in which, on the one hand, a genotype may arise with an auxin activity constantly higher than in a hermaphrodite and, on the other, the male genotype may arise with a constantly lower auxin activity, at least during the differentiation of the flower. A stable dioecious species requires the transformation of a pair of homologous chromosomes into a pair that is at any rate partly non-homologous.

The same monoecious population could give rise to the origin of female apomictic plants in case of an accidental accumulation of + genes. Thus apomixis and dioecism would be related to each other with regard to their origin.

Apomixis could even originate more easily as no special mechanism of linkage is required to produce a male genotype next to a female one. The fact that among apomictic species a remarkable number of poly-ploids and hybrids is present (SWANSON, 1958) could have a physi-ological ground. Both polyploidy and hybridisation promote vege-tative growth, so increase the auxin activity. Especially in populations with transitions between monoecism and dioecism such a sudden rise would suffice for the origin of apomicts.

In animals specimen are known where one species consists of sexual and asexual generations. So in the Aphid, *Tetraneura* dioecism and parthenogenesis go side by side (WHITE, 1954). The sexual generation consists of males with 13 chromosomes and females with 14 chromo-somes. The parthenogenetic females are genotypically similar to the sexual females. Environmental conditions decide which of both female types will arise. Under conditions that benefit growth the eggs develop into parthenogenetic females, under less favourable circumstances sexual females will appear.

One may even wonder whether a relation would exist between auxin activity of the host plant and the sex expression in the parasites. As a matter of fact it is no longer surprising that plant hormones could affect processes in animals and vice versa. Since an increasing amount of data point to the fact that hormones might directly interfere with gene action, and we know the processes of gene action in animals and plants to be similar. LÖVE & LÖVE (1945) have promoted the development of pistils and suppressed the development of stamina in *Melandrium* with the animal sex-hormones oestrone and oestradiol. With testosterone the reverse effect had been obtained. Oestrone has stimulated the growth of pea embryos in vitro, whereas testosterone has strongly inhibited this growth (HELMKAMP & BONNER, 1953). ZOLLIKOFER (1939) has noticed a promoting effect of oestrone on the germination of seed. The same author could strongly promote the formation of "bulbils" in *Poa alpina vivipara* with oestrone. In the non-viviparous *Poa alpina fructifera* no influence has been observed on the development of the inflorescence.

Interference of artificially administered animal sex hormones on the physiology of plant cells seems obvious. That in plants and in animals circumstances that promote the development of female sex organs, as some hormones do, are beneficial for growth in general, might be explained by the fact that a female gamete has to be formed in a milieu in plant or animal, where favourable circumstances occur for growth, even after fertilisation. A condition which is not required for the development of male gametes. On the contrary, it is conceivable that it is less favourable circumstances for vegetative growth that are required in order to keep the gametes small, a feature which will meet the conditions required in relation to the mobility of the male gametes.

In dioecious organisms these different requirements of the gametes might invoke a more favourable milieu for vegetative growth in females than in males.

DIOECISM IN PLANT KINGDOM AND ANIMAL KINGDOM

The striking difference in frequency of dioecism in the plant kingdom and in the animal kingdom is still under discussion. In the animal kingdom most species are dioecious whereas in the plant kingdom dioecism is rather rare.

The circumstance that polyploidy is common in plants and rare in animals has given rise to a hypothesis in which both phenomena are correlated. MULLER (1925) noticed the disturbing effect of polyploidisation on sex determination in *Drosophila* and therefore concludes that polyploidy interferes with the sex-determining mechanism, as a consequence of which dioecism could not become a common feature in plants.

WESTERGAARD (1958) observed that polyploidy in a number of dioecious plant species, with others in *Melandrium*, does not interfere with the sex-determining mechanism. He has concluded that MULLER's statement does not hold generally.

We are of the opinion that the two mechanisms which lead to the origin of polyploidy and dioecism are totally different mechanisms. However, when these two processes coincide, polyploidy might hamper the processes that lead to the origin of dioecism in plants and in animals.

As regards organisation animals are from the lowest groups on much more complicated than plants. Their ontogeny will require a much more complicated team-work of genes than the ontogeny of plants will. Therefore polyploidy in animals will generally disturb the balanced team-work of the genes, and consequently cause a lowered vitality of the resulting genotypes (WETTSTEIN, 1927). FANKHAUSER (1945) obtained polyploids, among others, of *Triturus*, which appeared to have a distinctly decreased viability, owing to the vital organs not having proportionally increased.

Plants have a simpler organisation and therefore can endure polyploidy better. Even this condition appears to be profitable. However, the manifold appearance of polyploidy (50% of the higher plants being polyploids, RIEGER, 1963, p. 92) will interfere with the building up of a balanced system, as required for the origin of dioecism. In a population where transitions between monoecism and dioecism are present, polyploidisation might lead to the origin of apomixis, as we have stated before.

In this connection it is of interest to note the circumstance that where in the animal kingdom polyploidy has been observed as in Protozoa, Crustaceae, Coleoptera, Lepidoptera and Orthoptera, it is practically always attended by parthenogenesis (SWANSON, 1958).

We would state that polyploidy is such an unfavourable condition in animals that it could not become of any importance in the animal kingdom. Whereas dioecism is favourable and could develop. In plants both polyploidy and dioecism are favourable. Polyploidy could evolve fairly easily and consequently delayed the spreading of dioecism over the plant kingdom.

The author would like to express his gratitude and great respect towards the late Professor Dr. C. L. RÜMKE who initiated and encouraged these investigations.

He wishes to thank Professor Dr. G. A. VAN ARKEL for his criticism and the many valuable discussions which considerably improved the presentation of this study. He is particularly indebted to Professor Dr. M. WESTERGAARD who suggested to repeat the investigations of CORRENS on pollen storage and supplied him with some plant material.

REFERENCES

ÅKERLUND, E. (1927). Ein *Melandrium* – Hermaphrodit mit weiblichem Bestand. *Hereditas* **10**: 153–159.

AUDUS, L. J. (1963). Plant Growth Substances. Leonard Hill, London. Interscience Publishers, New York.

BAKER, H. G. (1947). Infection of species of *Melandrium* by *Ustilago violacea* (Pers.) Fucke and the transmission of the resultant disease. *Ann. Bot. N.S.* **11**: 333–348.

BAKER, H. G. (1948). Stages in invasion and replacement demonstrated by species of *Melandrium*. *J. Ecol.* **36**: 96–119.

BEERMAN, W. & U. CLEVER (1964). Chromosome puffs. *Sc. American* **210** (4): 50–58.

BLACKBURN, K. B. (1923). Sex chromosomes in plants. *Nature* **112**: 687–688.

BLACKBURN, K. B. (1924). The cytological aspects of the determination of sex in the dioecious forms of *Lychnis*. *Brit. Journ. Exp. Biol.* **1**: 413–429.

BLACKBURN, K. B. (1929). On the occurrence of sex chromosomes in flowering plants with some suggestions as to their origin. *Proc. Internat. Congr. Plant Sci.*, I: 299–306.

CORRENS, C. (1918). Fortsetzung der Versuche zur experimenteller Verschiebung des Geschlechtsverhältnisses. *Sitzungsber. d. Preuss. Akad. d. Wiss. Phys.-Mat. Klasse* (1918): 1175–1200.

CORRENS, C. (1921a). Versuche bei Pflanzen das Geschlechtsverhältnis zu verschieben. *Hereditas* **2**: 1–24.

CORRENS, C. (1921b). Zweite Fortsetzung der Versuche zur experimentellen Verschieben des Geschlechtsverhältnisses. *Sitzungsber. d. Preuss. Akad. d. Wiss. Phys.-Mat. Klasse* (1921): 330–354.

CORRENS, C. (1922). Alkohol und Zahlenverhältnis der Geschlechter bei einer getrennt geschlechtigen Pflanze (*Melandrium*). *Naturw.* **49**: 1049–1052.

CORRENS, C. (1924). Über den Einflusz des Alters der Keimzellen. I. *Sitzungsber. d. Preuss. Akad. d. Wiss.* (1924): 70–104.

CRAFTS, A. S. (1961). Translocation in plants. Holt, Rinehart and Winston, New York.

DAVIDSON, E. H. (1965). Hormones and genes. *Sc. Amer.* **212** (6): 36–45.

EMERSON, R. A. (1932). The present status of maize genetics. *Proc. 6th Intern. Congr. Genet.*, I: 141–167.

FAGERLIND, F. (1946). Hormonale Substanzen als Ursache der Frucht- und Embryobildung bei pseudogamen *Hosta*-Biotypen. *Svensk Bot. Tidskr.* **40**: 230–234.

FANKHAUSER, G. (1945). The effect of changes in chromosome number on amphibian development. *Quart. Rev. Biol.* **20**: 20–78.

GALUN, E., Y. JUNG & A. LANG (1962). Culture and sex modification of male cucumber buds in vitro. *Nature* **197**: 596–598.

GOEBEL, K. (1908). Einleitung in die experimentelle Morphologie der Pflanzen. Teubner, Leipzig.

GOEBEL, K. (1913). Organographie der Pflanzen. Fischer, Jena.

GOLDSCHMIDT, R. B. (1955). Theoretical genetics. University of California Press, Berkeley and Los Angeles.

GREGORY, F. G. & J. A. VEALE (1957). A re-assessment of the problem of apical dominance. *Symp. Soc. Exp. Biol.* **11**: 1–20.

GRUBER, M. & R. N. CAMPAGNE (1965). Regulation of protein synthesis: An alternative to the repressor-operator hypothesis. *Proc. Kon. Ned. Akad. v. Wet.* Series C, **68** (4): 1–7.

GUSTAFSON, F. G. (1939). The cause of natural parthenocarpy. *Am. Journ. of Bot.* **26**: 135–138.

GUSTAFSON, A. (1946). Apomixis in the higher plants. I. The mechanism of apomixis. *Lunds Univ. Arsskr.* N. F. Avd. 2, **42**(3): 1–66.

GUSTAFSON, A. (1947a). Apomixis in the higher plants. II. The causal aspects of apomixis. *Lunds Univ. Arsskr.* N. F. Avd. 2, **43**(2): 71–178.

GUSTAFSON, A. (1947b). Apomixis in the higher plants. III. Biotype and species formation. *Lunds Univ. Arsskr.* N. F. Avd. 2, **43**(12): 183–370.

HAMILTON, T. H., R. J. MOORE, A. F. RUMSEY, A. R. MEANS & A. R. SCHRANK (1965). Stimulation of synthesis of ribonucleic acid in sub-apical sections of *Avena* coleoptile by indolyl-3-acetic acid. *Nature* **208**: 1180–1183.

HELMKAMP, G. & J. BONNER (1953). Some relationships of sterols to plant growth. *Plant Physiol.* **28**: 428–436.

HERTWIG, G. & P. HERTWIG (1922). Die Vererbung des Hermaphroditismus bei *Melandrium*. *Z.I.A.V.* **28**: 259–294.

HESLOP-HARRISON, J. (1956). Auxin and sexuality in *Cannabis sativa*. *Physiol. Plant.* **9**: 588–597.

HESLOP-HARRISON, J. (1957). The experimental modification of sex expression in flowering plants. *Biol. Rev. Cambridge Phil. Soc.* **32**: 38–90.

JACOB, F. & J. MONOD (1961). Genetic regulatory mechanisms in the synthesis of proteins. *J. Mol. Biol.* **3**: 318–356.

JONES, D. F. (1932). The interaction of specific genes determining sex in dioecious maize. *Proc. 6th Intern. Congr. Genet.*, II: 104–107.

JONES, D. F. (1934). Unisexual maize plants and their bearing on sex differentiation in other plants and animals. *Genetics* **19**: 552–567.

LAIBACH, F. & F. J. KRIBBEN (1950). Der Einfluss von Wuchsstoff auf die Bildung männlicher und weiblicher Blüten bei einer monözischen Pflanze (*Cucumis sativa* L.). *Ber. dtsch. Bot. Ges.* **62**: 53–55.

LAIBACH, F. & F. J. KRIBBEN (1951a). Der Einfluss von Wuchsstoff auf das Geschlecht der Blüten bei einer monözischen Pflanze. *Beitr. Biol. Pfl.* **28**: 64–67.

LAIBACH, F. & F. J. KRIBBEN (1951b). Die Bedeutung des Wuchsstoffs für die Bildung und Geschlechtsbestimmung der Blüten. *Beitr. Biol. Pfl.* **28**: 131–142.

LEWIS, D. (1942). The evolution of sex in flowering plants. *Biol. Rev. Cambridge Phil. Soc.* **17**: 46–67.

LIBBERT, E. (1954). Zur Frage nach der Natur der Korrelativen Hemmung. *Flora* **141**: 271–297.

LIBBERT, E. (1955). Nachweis und chemischer Trennung des Korrelationshemmstoffes und seiner Hemmstoffvorstufe. *Planta* **45**: 405–425.

LÖVE, D. (1942). Intersexuality in *Melandrium rubrum*, probably caused by a translocation between the sex chromosomes. *Hereditas* **28**: 497–498.

LÖVE, D. (1944). Cytogenetic studies on dioecious *Melandrium*. *Botan. Notiser* (1944): 125–214.

LÖVE, A. & D. LÖVE (1945). Experiments on the effects of animal sex hormones on dioecious plants. *Arkiv. Botan. Stockholm* **32**A: 1–60.

MITCHELL, W. D. & S. H. WITTWER (1962). Chemical regulation of flower sex expression and vegetative growth in *Cucumis sativus* L. *Science* **136**: 880–881.

MULLER, H. J. (1925). Why polyploidy is rarer in animals than in plants. *Am. Naturalist* **66**: 118–138.

NIGTEVECHT, G. VAN (1966). Genetic studies in dioecious *Melandrium*. I. Sexlinked and sex-influenced inheritance in *M. album* and *M. dioicum*. *Genetica* **37**: 281–306.

NITSCH, J. P., E. B. KURZ, J. L. LIVERMAN & F. W. WENT (1952). The development of sex expression in cucurbit flowers. *Am. J. Botany* **39**: 32–43.

NITSCH, J. P. (1965). In: Handbuch der Pflanzenphysiologie, XV: 1537–1647. Springer Verlag, Berlin.

NOODÉN, L. D. & K. V. THIMANN (1963). Evidence for a requirement for protein synthesis for auxin-induced cell enlargement. *P.N.A.S.* **50**: 194–200.

ÖSTERGREN, G. & W. K. HENEEN (1962). A squash technique for chromosome morphological studies. *Hereditas* **48**: 332–341.

OVERBEEK, J. VAN, M. E. CONKLIN & A. F. BLAKESLEE (1941). Chemical stimulation of ovule development and its possible relation to parthenogenesis. *Am. J. Botany* **28**: 647–656.

PETERSON, C. E. & L. D. ANHDER (1960). Induction of staminate flowers on gynoecious cucumbers with gibberellin A3. *Science* **131**: 1673–1674.

RIEGER, R. (1963). Die Genommutationen. Fischer, Jena.

SCHAFFNER, J. H. (1922). Control of sexual state in *Arisaema triphyllum* and *A. draconticum. Am. J. Botany* **9**: 72–78.

SCHAFFNER, J. H. (1925). Experiments with various plants to produce change of sex in the individual. *Bull. Torrey Bot. Cl.* **52**: 35–47.

SHIFRISS, O. (1961). Gibberellin as sex regulator in *Ricinus communis. Science* **133**: 2061–2062.

SHIFRISS, O. & W. L. GEORGE (1964). Sensitivity of female inbreds of *Cucumis sativus* to sex reversion by gibberellin. *Science* **143**: 1452–1453.

SHULL, G. H. (1910). Inheritance of sex in *Lychnis. Bot. Gaz.* **49**: 110–125.

SHULL, G. H. (1911). Reversible sex-mutants in *Lychnis dioica. Bot. Gaz.* **52**: 329–368.

SNOW, R. (1940). A hormone for correlative inhibition. *New Phytol.* **39**: 177–184.

STEBBINS, G. L. (1951). Variation and evolution in plants. Columbia University Press, New York.

SWANSON, C. P. (1958). Cytology and cytogenetics. MacMillan, London.

THIMANN, K. V. (1954). Correlations of growth by humoral influences. *VIIIe Congr. Internat. Bot. Rapp. et Comm. Sect.*, XI: 114–128.

THIMANN, K. V. & F. SKOOG (1934). On the inhibition of bud development and other functions of growth substances in *Vicia faba. Proc. Roy. Soc.* B **114**: 317–339.

TIEDJENS, V. A. (1928). Sex ratio in cucumber flowers as affected by different conditions of soil and light. *Journ. Agr. Res.* **36**: 721–746.

VARNER, J. E. & G. R. CHANDRA (1964). Hormonal control of enzyme synthesis in barley endosperm. *P.N.A.S.* **52**: 100–106.

VARNER, J. E. & G. R. CHANDRA (1965). Gibberellic acid-controlled metabolism of RNA in aleurone cells of Barley. *Biochim. et Biophys. Acta* **108** (4): 583–592.

WARMKE, H. E. & A. F. BLAKESLEE (1939). Effect of polyploidy upon the sex mechanism in dioecious plants. *Genetics* **24**: 88–89.

WARMKE, H. E. (1946). Sex determination and sex balance in *Melandrium. Am. J. Botany* **33**: 648–660.

WESTERGAARD, M. (1940). Studies on cytology and sex determination in polyploid forms of *Melandrium album. Dansk. botan. Arkiv.* **10**: 1–131.

WESTERGAARD, M. (1946). Aberrant Y-chromosomes and sex expression in *Melandrium album. Hereditas* **32**: 419–443.

WESTERGAARD, M. (1948). The relation between chromosome constitution and sex in the offspring of triploid *Melandrium. Hereditas* **34**: 257–279.

WESTERGAARD, M. (1958). The mechanism of sex determination in dioecious flowering plants. *Adv. in Genetics* **9**: 217–281.

WETTSTEIN, F. VON (1927). Die Erscheinung der Heteroploidie, besonders im Pflanzenreich. *Ergebn. der Biologie* **2**: 311–356.

WHITE, M. J. D. (1954). Animal Cytology and Evolution (2nd ed). Cambridge Univ. Press, New York.

WINGE, Ø. (1923). On sex chromosomes, sex determination and preponderance of females in some dioecious plants. *Compt. rend. trav. lab. Carlsberg* **15**: 1–26.

WINGE, Ø. (1931). X- and Y-linked inheritance in *Melandrium*. *Hereditas* **15**: 127–165.

YAMPOLSKY, C. & H. YAMPOLSKY (1922). Distribution of sex forms in the phanerogamic flora. *Bibliotheca Genet.* **3**: 1–62.

YOUNG, J. O. (1943). Histological comparison of cucumber fruits developing parthenocarpically and following pollination. *Bot. Gaz.* **105**: 69–79.

ZOLLIKOFER, C. (1939). Der Einfluss tierischer Hormone auf Pflanzen. *Ber. Schweiz. Bot. Ges.* **49**: 427–428.

CURRICULUM VITAE

De auteur werd op 15 maart 1930 te Hilversum geboren. Na het behalen van het diploma H.B.S. B aan de Gemeentelijke H.B.S. te Hilversum ving hij in 1947 te Utrecht zijn studie aan in de biologie. In 1951 deed hij candidaatsexamen en in 1954 doctoraalexamen biologie met hoofdrichting plantensystematiek, nevenrichting plantenphysiologie, bijvakken algemene zoologie en phytopathologie. Van 1952–1955 was hij assistent aan het Instituut voor Systematische Plantkunde te Utrecht. Sindsdien is hij verbonden aan het Genetisch Instituut te Utrecht, sedert 1964 als wetenschappelijk hoofdambtenaar. Gedurende deze laatste periode bracht hij tweemaal een werkbezoek van ruim 3 maanden aan het Genetisch Instituut te Lund (Zweden) waar hij bij Professor Dr. G. Östergren en Dr. A. Lima de Faria cytogenetisch onderzoek verrichtte.